MATHEMATICS RESEARCH DEVELOPMENTS

MATHEMATICAL MODELING

METHODS, APPLICATIONS AND RESEARCH

MATHEMATICS RESEARCH DEVELOPMENTS

Additional books in this series can be found on Nova's website under the Series tab.

Additional e-books in this series can be found on Nova's website under the eBooks tab.

MATHEMATICS RESEARCH DEVELOPMENTS

MATHEMATICAL MODELING

METHODS, APPLICATIONS AND RESEARCH

SETH SPARKS
AND
BILL WILLIS
EDITORS

nova
science publishers
New York

Additional color graphics may be available in the e-book version of this book.

Library of Congress Cataloging-in-Publication Data

ISBN: 978-1-53613-162-8

Published by Nova Science Publishers, Inc. † New York

CONTENTS

PREFACE

Mathematical Modeling: Methods, Applications and Research reviews recent progress in three different propulsion systems which may offer significant advantages for more efficient, compact and adaptive astronautic vehicles in the mid-21st century: swirling nuclear magneto-hydrodynamic propulsion, biomimetic magneto-rheological propulsion and electro-hydrodynamic propulsion systems. The authors introduce an empirical mathematical model. This model has been developed to describe contrast uptake and washout behavior without use of vascular input functions. Though this approach does not require making assumptions about underlying physiology or anatomy, the primary disadvantage of this approach, is that the parameters obtained by this approach do not correspond directly to identifiable physiological or anatomic features. In closing, the authors aim to express how the mathematical modeling is a valid tool for teaching and learning in a competent way and should be incorporated in academic curricula.

Chapter 1 - The long-standing interest in manned inter-planetary missions has motivated a re-think worldwide in developing better propulsion drives. Many hypothetical systems which did offer some promise in the 1970s and 1980s have been abandoned. A significant re-surge has occurred in electromagnetic propulsion (EMP) systems and a new one in biomimetic systems. The general consensus now is that chemical rocket propulsion simply cannot achieve the tremendous distances involved in manned missions to other planets, to say nothing of inter-stellar travel. In this paper, the authors review recent progress in three different propulsion systems which may offer significant advantages for more efficient, compact and adaptive astronautical vehicles in the mid-21st century. Swirling nuclear magneto-hydrodynamic propulsion, biomimetic magneto-rheological propulsion and electro-hydrodynamic propulsion systems are explored. The fluid mechanics, heat transfer and electromagnetics of these systems is addressed via nonlinear boundary value problems. These provide a good perspective of the parameters controlling efficiency of propulsion. A diverse range of numerical codes (developed by the first author) are applied to solve these boundary value problems. These include the differential transform method (DTM), numerical integration scheme (NIS) and Chebyschev spectral collocation method (SCM). Other proposed EHD and MHD propulsion systems are also reviewed briefly. Another focus of this chapter is to promote the use of the described numerical methods in future studies of EMP, as a compliment to conventional computational methods such as the finite element method (FEM) and computational fluid dynamics (CFD) software e.g., FLUENT (which although capable of complex geometric modelling often neglect important multi-physical flow characteristics).

Chapter 2 - A tracer kinetic model is a mathematical description of the transport processes of the tracer within the tissue. Various methods have been proposed for the quantification of kinetic parameters including perfusion. These methods can be roughly classified into model-dependent and model-independent methods. The model-dependent method usually employs compartment models and generally requires specific knowledge about the vascular and compartmental architecture. The kinetic parameters are generally estimated using the nonlinear least-squares (NLSQ) method. Although the NLSQ method is a powerful tool for estimating kinetic parameters, it is not efficient in terms of computation time. Furthermore, poor initial guesses result in long computation times or occasionally failure to converge on a good fit. As an alternative approach, the linear least-squares (LLSQ) method has been introduced, which is computationally efficient and does not need initial guesses of the parameters. The graphical method has also been proposed. The model-independent method yields kinetic parameters for a tissue region described as a "black box" and thus does not provide insight into its internal structure. For quantification of kinetic parameters using the model-independent method, deconvolution between the measured tissue residue curve and the vascular input function is needed to obtain the tissue response function. Deconvolution, however, is an ill-posed problem especially when there exists statistical noise. Thus, it is necessary to convert an ill-posed problem to a well-posed one using the so-called "regularization". Of the regularization methods, the truncated singular value decomposition (TSVD) is a popular method for obtaining regularized solutions. In this chapter, the use of TSVD in the model-independent method is introduced. The application of spectral analysis is also introduced. This method provides a spectrum of the kinetic components which are involved in the regional uptake and partitioning of tracer from the blood to the tissue and allows the tissue response function to be derived with minimal modeling assumptions. The kinetic models employed in compartment analysis can be classified into lumped parameter and distributed parameter models. In the lumped parameter model, the compartments are assumed to be well-mixed, that is, tracer movement is sufficiently fast and distributes evenly throughout the compartments, such that tracer concentration is only a function of time, but not space. On the other hand, a distributed parameter model attempts to account for tracer concentration gradient within the compartments, such that concentration can be a function of both time and space. One of the simple distributed parameter models is the tissue homogeneity model. In this chapter, adiabatic approximation to the tissue homogeneity model is also described. Finally, an empirical mathematical model is introduced. This model has been developed to describe contrast uptake and washout behavior without use of vascular input functions. This approach does not require making assumptions about underlying physiology or anatomy. The primary disadvantage of this approach, however, is that the parameters obtained by this approach do not correspond directly to identifiable physiological or anatomic features. This chapter will be helpful for gaining a deeper understanding of mathematical modeling of tracer kinetics for medical applications.

Chapter 3 – The author's goal is to establish a relation between Mathematics and reality. Actually, from daily matters, the author is demonstrating that these situations can be interpreted building Mathematical Models which bring information about all of them, in this case Algebra of Matrices. Firstly, a matrix is introduced as a Mathematical Model of an image and then, the subtraction or the difference of matrices shows a new way to compare images. To reach this model is necessary a specific software (Octave), consequently the digital competence become part of this work. Secondly, naturally the concept of the Scaled Euclidean Product has been developed, also the product of matrices as Mathematical Models of Matrix Algebra for daily situations related to Supermarket shopping. Consequently, these daily situations lead to introduce Matrix Models linking quantities, prices and bought products; a concept which allows a transverse and interdisciplinary branch of the Mathematics. With the proposed examples is possible to assure that the Mathematical Modeling is a valid tool for teaching / learning in a competent way, at the same time recommendable to be included in the academic curricula, reaching a transverse view of the Mathematics Actually, it materializes epistemological, heuristic and cognitive aspects in the context of the Mathematics teaching. Synthesizing the authors can consider mathematical modeling as an efficient teaching innovation tool, and a transmission strap that provides the knowledge approaching mathematics to the reality.

In: Mathematical Modeling ISBN: 978-1-53613-162-8
Editors: Seth Sparks and Bill Willis

Chapter 1

MULTI-PHYSICAL ELECTRO-MAGNETIC PROPULSION FLUID DYNAMICS: MATHEMATICAL MODELLING AND COMPUTATION

***O. Anwar Bég*[*], PhD**
PGCAP FHEA SMAIAA, Aeronautical and Mechanical Engineering,
University of Salford, Manchester, UK

ABSTRACT

The long-standing interest in manned inter-planetary missions has motivated a re-think worldwide in developing better propulsion drives. Many hypothetical systems which did offer some promise in the 1970s and 1980s have been abandoned. A significant re-surge has occurred in electromagnetic propulsion (EMP) systems and a new one in biomimetic systems. The general consensus now is that chemical rocket propulsion simply cannot achieve the tremendous distances involved in manned missions to other planets, to say nothing of inter-stellar travel. In this paper, we review recent progress in three different propulsion systems

[*] Corresponding Author Email: O.A.Beg@salford.ac.uk.

which may offer significant advantages for more efficient, compact and adaptive astronautical vehicles in the mid-21st century. Swirling nuclear magneto-hydrodynamic propulsion, biomimetic magneto-rheological propulsion and electro-hydrodynamic propulsion systems are explored. The fluid mechanics, heat transfer and electromagnetics of these systems is addressed via nonlinear boundary value problems. These provide a good perspective of the parameters controlling efficiency of propulsion. A diverse range of numerical codes (developed by the first author) are applied to solve these boundary value problems. These include the differential transform method (DTM), numerical integration scheme (NIS) and Chebyschev spectral collocation method (SCM). Other proposed EHD and MHD propulsion systems are also reviewed briefly. Another focus of this chapter is to promote the use of the described numerical methods in future studies of EMP, as a compliment to conventional computational methods such as the finite element method (FEM) and computational fluid dynamics (CFD) software e.g., FLUENT (which although capable of complex geometric modelling often neglect important multi-physical flow characteristics).

Keywords: numerical simulation, Maxwell electromagnetic equations, electrohydrodynamics, ions, Lorentz force, Chebyschev spectral collocation (CSC), differential transform method (DTM), Mathematica

1. INTRODUCTION

Electromagnetic propulsion (**EMP**) systems have been studied for many years. A lucid summary of progress up to 1968 has been given in the tremendous monograph by Princeton university aerospace engineer Robert Jahn [1]. A further summary of progress in the ensuing three decades has been provided by Jahn and Choueiri [2]. This form of propulsion encompass a broad variety of strategies for achieving very high exhaust velocities in order to reduce the total propellant burden and corresponding launch mass of present and future space transportation systems. These techniques group broadly into three categories:

- Electrothermal propulsion: the propellant is electrically heated, then expanded thermodynamically through a nozzle.

- Electrostatic and Electrohydrodynamic propulsion: wherein ionized propellant particles are accelerated through an electric field.
- Electromagnetic propulsion: wherein current driven through a propellant plasma interacts with an internal or external magnetic field to provide a stream-wise body force.

All three systems can produce a range of exhaust velocities and payload mass fractions an order of magnitude higher than that of the most advanced chemical rockets, which can thereby enable or substantially enhance many attractive space missions. We consider only the second and third type here. The attainable thrust densities (thrust per unit exhaust area) of these systems are much lower, however, which predicates longer flight times and more complex mission trajectories. In addition, these systems require space-borne electric power supplies of low specific mass and high reliability, interfaced with suitable power processing equipment.

Optimization of electromagnetic propulsion systems thus involves multidimensional trade-offs among mission objectives, propellant and power plant mass, trip time, internal and external environmental factors, and overall system reliability. An enduring international program of research and development of viable electro/magnetic thrusters has been in progress for several decades, and over the past few years this has led to the increasing use of a number of EMP systems on commercial and governmental spacecraft. The NASA Xenon thruster is a key development in this regard [3]. Over 200 spacecraft operated throughout the solar system have used some form of EMP whether for station-keeping, orbit raising, or primary propulsion.

A key target in terms inter-planetary missions is the attainment of much higher specific impulse e.g., 100 km/s. This is however totally inadequate for interstellar journeys which involve colossal distances and frequently very severe (in many cases uncharted) environments. Solar and laser powered electro-rocket with an external power source seem the best proposition for interstellar missions.

Although electric propulsion is not appropriate for terrestrial launches from the Earth's surface, owing to weak average thrust, in space, it may be the best proposition for humanity. Many investigations have been communicated on the overall performance of different EMP thrusters. They have however ignored key fluid dynamical aspects. In this chapter, which serves as a complement to the others in this book, rather than examining thruster designs, performance criteria and construction, we dwell on the multi-physical fluid dynamics, thermophysics and electro-magneto-dynamics of three different systems. Many concepts previously not considered are described including entropy generation minimization, biomimetic sustainable propulsion with Lorentzian forces for different working fluids (the example of a robust electrically-conducting gel which is non-Newtonian is considered) and electrohydrodynamic ion drag systems.

2. Electrohydrodynamic Thrusters

Electrohydrodynamics (EHD), or alternatively electro-fluid-dynamics (EFD) or electrokinetics, refers to motion of electrically charged fluids under the influence of applied electric fields. The electrical engineer, Russel Melcher of MIT [4] has given the best exposition to date of this subject. The articles of English physicist, G. I. Taylor [5, 6], are also useful, although applications therein are focused on atmospheric phenomenon e.g., thunderstorm generation, rather than engineering propulsion. In EHD-thrusters the applied electric field accelerates the electrically charged particles while collisional coupling between the charged particles and neutral particles is used to convert the directed energy of the charged particles into neutral gas/fluid particle flow without the need for moving parts, Figure 1.

Since the momentum transfer between ions and neutral particles is much more efficient than between electrons and neutral particles, neutral gas flows produced in EHD-devices are commonly referred to as ion wind (also ionic wind, ion-driven wind, corona wind, electric wind, and others) flows. EHD flow is the flow of neutral particles caused by the drifting of ions in an electric field. In many cases these ions are generated by a positive high voltage corona discharge in the high curvature portions of the electrodes. The corona wire has a uniform high curvature and therefore will generate ions uniformly. The positively ionized gas molecules will travel from the corona wire ion source towards the collector (ground) colliding with neutral molecules in the process. These collisions will impart momentum to the neutral atoms that will move towards the collector as a result. The momentum gained by the neutral gas is exactly equal to the momentum gained by the positive ions that move through the electric field between the electrodes, and lost in inelastic collisions to the neutral molecules.

Corona discharges are non-equilibrium plasmas with an extremely low degree of ionization. There exist two zones with different properties, the ionization zone and the drift zone. The energy from the electric field is mainly absorbed by electrons in the ionization zone, immediately close to the corona electrode, and transmitted to the neutral gas molecules by inelastic collisions producing electron positive ion pairs, where the net space charge density will remain approximately zero. Since there is a much higher mobility of electrons relative to the positive ions, and because the only region were (positive) ions will be generated is in the ionization region where the electrons have enough energy to accomplish that due to the much higher electric field intensity. Also, in the drift region the electrons and ions do not have enough energy to react with neutrals and also have too low density to react with other ionized particles.

These physical aspects can be very much simplified for simulating EHD propulsion. A schematic of a robust EHD thruster is given in Figure 1. The ions are “seeded” at the anode and are absorbed at the cathode. Here we do not consider the mechanism for creating the ions. They can be created, for example, by corona discharge, or by other means. The neutral gas flow is coming from the anode side of the thruster. Due to the friction forces with ions, the neutral gas flow is accelerated in the discharge thruster region that creates the thrust, as in an ordinary air turbine jet. We dwell on the laminar flow case only. Electrostrictive forces are ignored. Electrical field distribution is taken as *uni-directional* and parallel to the flow direction with unipolar ions present in the fluid and mobility constant. The emitter and collector electrodes do not influence the fluid flow. Ion current and field distributions are also unaffected by surface charges on the insulating boundaries of the conduit.

We further assume, following [7] that *ion velocity* relative to *fluid velocity* is of the same order of magnitude as the fluid velocity to the electrodes. High ion density exists in the medium. The conservation laws for electrical field, current density and electrical potential take the form:

2.1. Gauss's Law

$$\nabla \cdot \varepsilon \vec{E} = \rho \tag{1}$$

2.2. Conservation of Charge

$$\nabla \cdot \vec{J} + \frac{\partial \rho}{\partial t} = 0 \tag{2}$$

2.3. Conservation of Electric Potential

$$\vec{E} = -\nabla \Phi \tag{3}$$

When all three mechanisms of the current flow are combined we obtain the following relation for *total current density*:

$$\vec{J} = \sigma \vec{E} + \rho \vec{u} + \rho \mu \vec{E} \tag{4}$$

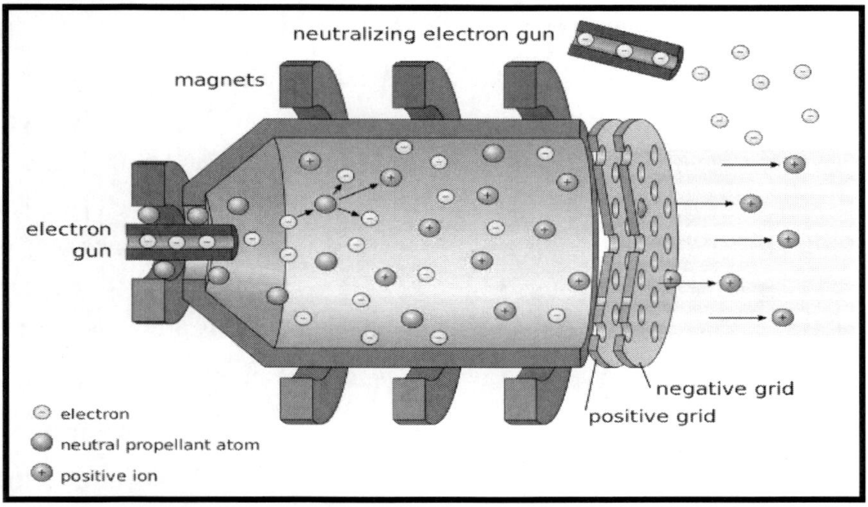

Figure 1. EHD thruster schematic.

We consider unipolar DC EHD pump flow in an EHD thruster system. The regime can be simplified to that shown in Figure 2.

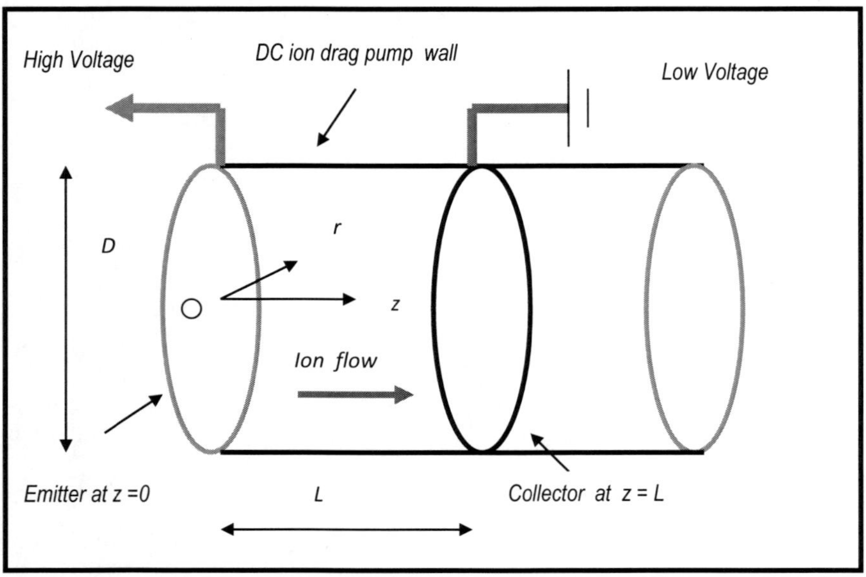

Figure 2. DC Ion drag propulsion.

where ∇ is the gradient operator, $\vec{E}$ is electrical field vector (V/m^2), $\vec{J}$ is the current density vector (Amps/m^2), ρ is charge density (Coulomb/m^3), t is time, ε is permittivity (Faradays/m), Φ is potential (V), σ is electrical conductivity (S/m), $\vec{u}$ is fluid bulk velocity vector (m/s) and μ is ion mobility (m^2/Vs). Herein we shall study the situation where the electrical field located at emitter (i.e., at $z = 0$ in Figure 2), is *zero*. This is known as the *maximum pressure scenario* and has been considered recently by Bég et al. [8]. In this study we further extend the computations presented in [8]. In equation (4), $\sigma\vec{E}$ is the *electrical conduction* term, $\rho\vec{u}$ is the *electrical convection* term and $\rho\mu\vec{E}$ is the *ionic mobility* term. The conservation equations can be shown to reduce for constant property EHD flow, to the following group of coupled, nonlinear, ordinary differential equations, in which z i.e., axial coordinate along the axis of the dc pump, is the only independent variable:

$$(\mu E + u)\frac{d\rho}{dz} + \frac{\sigma}{\varepsilon}\rho + \frac{\mu}{\varepsilon}\rho^2 = 0 \tag{5}$$

$$\frac{dE}{dz} - \frac{\rho}{\varepsilon} = 0 \tag{6}$$

$$E + \frac{d\Phi}{dz} = 0 \tag{7}$$

The relevant boundary conditions are:

$$z = 0(emitter): \quad E = 0;\ \rho = \rho_o;\ \Phi = V_o \tag{8}$$

Another important case, namely the *Operating Condition Scenario* can also be studied wherein electrical potential (Φ*) at the *collector* ($z = L$) is instead specified, rather than at the emitter. Normalization of the eqns. (5) to (7) is achieved with the following dimensionless variables:

$$E^* = E\frac{L}{V_o}, \qquad \rho^* = \frac{\rho}{\rho_o}, \qquad \Phi^* = \frac{\Phi}{V_0},\ z^* = \frac{z}{L},$$

$$\mathrm{Re}_E = \frac{\varepsilon u}{\sigma L}, E_{sl} = \frac{\frac{\mu V_0}{L}}{u},\ E_s = \frac{\rho_0 L^2}{\varepsilon V_0} \tag{9}$$

The non-dimensional conservation equations for electrical field (E^*), potential (Φ^*), and charge density (ρ^*), thereby reduce to:

$$(1 + E_{sl}E^*)\frac{d\rho^*}{dz^*} + \frac{\rho^*}{\mathrm{Re}_E} + E_s E_{sl}\rho^{*2} = 0 \tag{10}$$

$$\frac{dE^*}{dz^*} - E_s\rho^* = 0 \tag{11}$$

$$E^* + \frac{d\Phi^*}{dz^*} = 0 \tag{12}$$

The corresponding transformed boundary conditions are:

$$At\ z^* = 0(emitter):\quad E^* = 0;\ \rho^* = 1;\ \Phi^* = 1 \tag{13}$$

In the eqns. (10) to (12), Re_E is the electrical Reynolds number (symbolizing the relative significance of the charge convection effect and defined as the ratio of the timescales of charge convection by flow to that for charge relaxation by ohmic conduction), E_{sl} is the electrical slip number (which provides an estimate of the relative motion of the injected charges with regard to the bulk fluid velocity, u). and E_s is electrical source number (based on the analogy between the *Fourier thermal conduction* equation and *Poisson's electrostatics* equation). For engineering applications, it is pertinent to study the pressure generation (ΔP) and power efficiency (η_{EHD}). These are defined respectively, ignoring shear stresses along the conduit walls, as follows [**8**]:

$$\Delta P = \frac{1}{2}\varepsilon E_L^{\ 2} \tag{14}$$

$$\eta_{EHD} = \frac{power\ \ output}{power\ \ input} = \frac{(uA)\Delta P}{[V_0 - V_L]I} \tag{15}$$

where ΔP denotes pressure generation of the pump (N/m^2), E_L is the electrical field at the collector electrode (V/m), E_0 is the electrical field at the emitter electrode, A is the cross-sectional area of the unipolar dc EHD pump (m^2), V_0 is the voltage applied to the emitter electrode (V), V_L is the voltage applied to the collector electrode (V), I is current (Amps), L is distance between the electrode pair (m). As the electrical field becomes zero at the emitter electrode, the electrical current is therefore:

$$I = JA = \rho_0 uA \tag{16}$$

where ρ_0 is the charge density at the emitter electrode (Colulomb/m^3). It follows that the EHD efficiency is:

$$\eta = \frac{\frac{1}{2}\varepsilon E_L^{\ 2}}{[V_0 - V_L]\rho_0} \tag{17}$$

In dimensionless form we have:

$$\eta = \frac{\frac{1}{2}E*_L^{\ 2}}{Es[1 - \Phi *_L]} \tag{18}$$

where $E*_L$ is the dimensionless electrical field at the collector electrode and $\Phi *_L$ is the dimensionless potential at the collector electrode. The expression (18) associates η_{EHD}, directly with the electrical field values at the emitter and the collector electrodes. The coupled boundary value problem defined by eqns. (10) to (12) under conditions (13) is also solved subsequently by the Chebyschev Spectral Collocation (SCM) numerical approach. Spectral methods [9, 10] have demonstrated enhanced accuracy and stability in recent years. These methods have been applied extensively in diverse areas of engineering simulation including spinning disk flows [11], tribological flows [12], non-Newtonian flows [13], gas dynamics [14], acoustics [15], physiological fluid dynamics [16], droplet combustion [17], radiation heat transfer [18] and geomechanics [19]. Bég and co-workers have also applied spectral methods to bio-magneto-fluid dynamic simulations of relevance to astronautics during re-entry [20, 21] and rotating duct heat transfer [22] and gel propulsion heat transfer [23]. To solve the electrohydrodynamic boundary value problems described in section 2, we employ a spectral collocation method of the *Chebyshev* type. We look for an approximate solution, which is a global Chebyshev polynomial of degree N defined on the interval [-1, 1]. We discretize the interval by using collocation points to define the Chebyshev nodes in [-1, 1], namely

$$x_j = \cos(\frac{j\pi}{N}), \qquad j = 0.1,2,.....N.. \tag{19}$$

The derivatives of the functions at the collocation points are given by:

$$f^n(x_j) = \sum_{j=0}^{N} d_{kj}^n f(x_j), \qquad n = 1,2. \tag{20}$$

where d_{kj}^n represents the differentiation matrix of order n and are given by

$$d^1{}_{kj} = \frac{4\gamma_j}{N} \sum_{n=0}^{N} \sum_{l=0,\, n+l=odd}^{n-1} \frac{n\gamma_n}{c_l} T_l^n(x_k) T_n(x_j), \qquad k, j = 0,1,....N, \tag{21}$$

$$d^2{}_{kj} = \frac{2\gamma_j}{N} \sum_{n=0}^{N} \sum_{l=0,\, n+l=even}^{n-2} \frac{n\gamma_n (n^2 - l^2)}{c_1} T_l^{\,n}(x_k) T_n(x_j), \qquad k, j = 0,1,....N, \tag{22}$$

where $T_n(x_j)$ are the Chebyshev polynomial and the coefficients γ_j and c_l are defined as:

$$T_n(x_j) = \cos(n \cos^{-1} x_j), \qquad \gamma_j = \begin{cases} \frac{1}{2} & j = 0, or\ N \\ 1 & j = 1,2,....N-1 \end{cases} \qquad c_l = \begin{cases} 2 & l = 0, or\ N \\ 1 & l = 1,2,....N-1 \end{cases} \tag{23}$$

As described above the Chebyshev polynomials are defined on the *finite* interval [-1,1]. Therefore to apply Chebyshev spectral method to our equation (7), we make a suitable linear transformation and transform the physical domain [0, 1] to Chebyshev computational domain [-1, 1]. We sample the unknown function *w* at the Chebyshev points to obtain the data vector $w = [w(x_0), w(x_1), w(x_2), ...w(x_N),]^T$. The next step is to find a Chebyshev polynomial P of degree N that interpolates the data, i.e., $P(x_j) = w_j,\ \ j = 0,1,...N,$ and obtain the spectral derivative vector w by differentiating P and evaluating at the grid points, i.e., $w'_j = P'(x_j),\ \ j = 0,1,...N.$ This transforms the nonlinear differential equation into a system nonlinear algebraic equations which are solved by Newton's iterative method starting with an initial guess. Similarly, we can apply ***CSCM*** to the coupled system of equations and boundary conditions (3.10)-(3.12), to generate the following system of algebraic equations.

$$2(1 + E_{sl} E^*) \sum_{j=0}^{N} d^1_{kj} \rho^*(x_j) + \frac{1}{\mathrm{Re}_E} \rho^*(x_j) + E_S E_{Sl} \rho^{*2}(x_j) = 0 \tag{24}$$

$$2\sum_{j=0}^{N} d_{kj}^{1} E^{*}(x_j) - E_s \rho^{*} = 0 \tag{25}$$

$$2\sum_{j=0}^{N} d_{kj}^{1} \Phi^{*}(x_j) + E^{*} = 0 \tag{26}$$

We solve nonlinear equation for ρ^{*} given by (24) using Newton's iterative method starting with initial guesses for $\rho^{*}, E^{*}, \Phi^{*}$, and the result of (24) can be subsequently used to solve (25) and (26). Numerical results for $\rho^{*}, E^{*}, \Phi^{*}$, are then obtained in an iterative way and elaborated next.

The results obtained with **SCM** are noticed to compare well with the benchmark results in the literature. We note that in the mathematical model, we have employed an electrical Reynolds number. An alternative is the electrical Hartmann number (Ha_e) wherein increasing values of this parameter correspond to a reduction in the contribution of the *viscous* force in the regime and an accentuation in the *electrical Coulomb force.* To validate the present **SCM** computations, we have reproduced two sets of profiles using the same set of parameters as employed by Seyed-Yagoobi et al. [24] using shooting quadrature and Bég et al. [25] using the PSPICE electro-thermal network simulation method. Spectral methods can provide exponential convergence in space and have been widely used to solve nonlinear fluid mechanics problems. Although these methods are relatively infrequently applied to simulate multi-dimensional complex geometries, for simple geometries e.g., channels, they provide excellent accuracy [26]. The dimensionless charge distributions and electrical field profiles for different E_{sl} are shown in Figures 3, 4 respectively. The profiles achieved with **SCM** correlate closely with the earlier solutions in [24] and [25] again testifying to the excellent accuracy of **SCM**.

In Figure 3 the charge density (ρ*) in the flow between the two electrodes (collector and emitter) indicates that for constant and very low electrical Reynolds number, *ReE* of 2, and *Es* = 1, the charge density (ρ*)

is decreased substantially with increasing electrical slip number (*Esl*). The *electrical slip number* provides an estimate of the relative motion of the injected charges with regard to the bulk fluid velocity (u). The E_{sl} = 0 case implies *zero contribution of the mobility component to the current density*. Electrical slip number exerts a more prominent role in controlling charge density distribution than the charge relaxation time (ε/σ) which facilitates the transport of charges in the ion flow to reach the collector electrode. The distributions computed in Figure 3, also correlate well the study by Singhal and Garimella [27]. As specified by the particular boundary conditions imposed, charge density is maximized at the entry to the conduit (emitter) and decay gradually to zero at the collector (z^* = 1). Inspection of Figure 4 shows that electrical field, E^* is significantly lowered with an increase in E_{sl}. $E_{sl} = \frac{\frac{\mu V_0}{L}}{u}$ and is strongly enhanced with mobility and decreased with bulk fluid velocity (in any scenario, emitter potential (V_0) and electrode pair separation (L) will be invariant). It is important to emphasize that the flow scenario is based on ordinary *bipolar conduction* where *molecular dissociation* within the fluid, as opposed to injection from the electrodes, generates an equal number of positive and negative charges. The resultant current path assists strongly the power requirements for the direct current unipolar EHD pump. The transport of net space charges within the moving fluid moving is represented by the current density relation which features both a convection term, and an ionic mobility term.

Higher ion drag pump efficiency is generally achieved with *lower* ionic mobilities. To maximize pumping efficiencies, low E_{sl} values are necessary, and these are often ensured via working fluids with a high dielectric constant and low viscosity. When electrical slip effects are absent (in E_{sl} =0) the electrical field profile is infact approximately linear; this pattern is strongly altered as electrical slip effect is enhanced and we observe that for E_{sl} =10 (maximum considered), the profile clearly exhibits the greatest montonicity. Figure 5 shows the effects of E_{sl} on charge density variation with a much lower electrical Reynolds number (Re_E = 50, compared with 200 in Figure 3) and a lower electrical source number (Es = 0.7, compared with 1.0 in Figure 3). Considerably higher values of charge density are sustained throughout the channel. This indicates that electrical Reynolds number also exerts a strong *opposing* effect to charge transport in the regime as does the electrical source effect. We note further that with lower E_{sl} values as studied in Figure 5 (compared with Figure 3), a much stronger charge distribution is achieved. Figure 6 depicts the electrical field distribution with significantly lower E_{sl} values but a much higher electrical Reynolds number ($Re_E = 50$ compared with 2 in Figure 4).

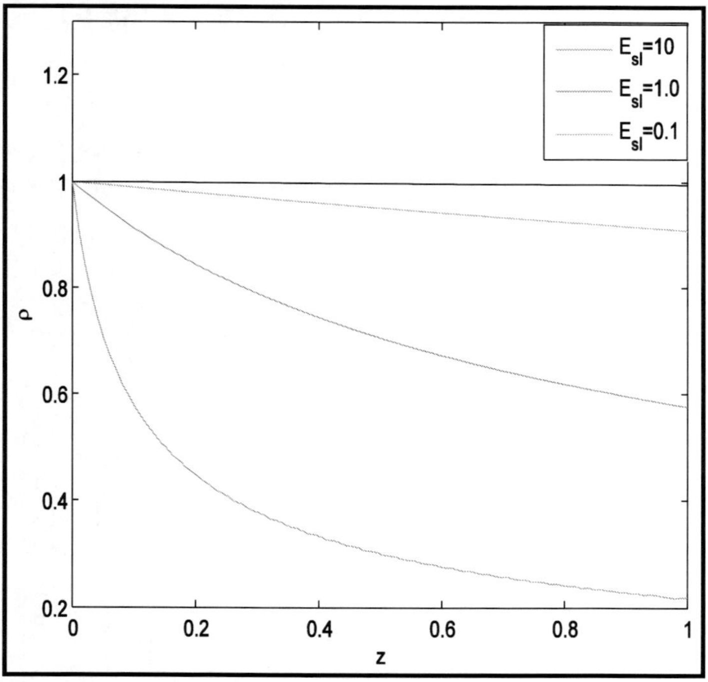

Figure 3. Distribution of charge density for $Re_E = 200$, $Es = 1$ for various E_{sl} values.

The benefit of much lower electrical slip number on electrical field growth through the conduit is again confirmed with Figure 6. Figure 7 shows that electric potential variation (Φ^*) is enhanced with electrical slip number increasing, particularly with closer proximity to the collector electrode ($z = 1$). For any slip number we observe that profiles are generally monotonic decays from the maximum at the emitter electrode to the collector (minimum). The influence of the electrical source number, E_s, on charge distribution and electrical potential evolution in the conduit are shown, respectively in Figures 8, 9. An increase in the electrical source number strongly depresses the charge distribution values, as shown in Figure 8. E_s is based on the analogy between *thermal conduction* (Fourier's equation) and *electrostatics* (Poisson's equation). The *ratio of heat source (W/m^3) to thermal conductivity (W/mK)* i.e., q'/k in Fourier's law is analogous to the *ratio of space charge density to electrical permittivity (ρ/ε)* in Poisson's equation, as elaborated by Melcher [4]. By assuming a uniform charge distribution, E_s relates *directly* to the space charge limitations identified in previous studies [24]. Electrical potential (Figure 9) is also decreased with increasing electrical source number; however unlike the charge distribution response, with high E_s numbers, the potential decays much more quickly to zero, along the conduit length. The point of vanishing potential is progressively displaced closer to the conduit emitter with increasing E_s.

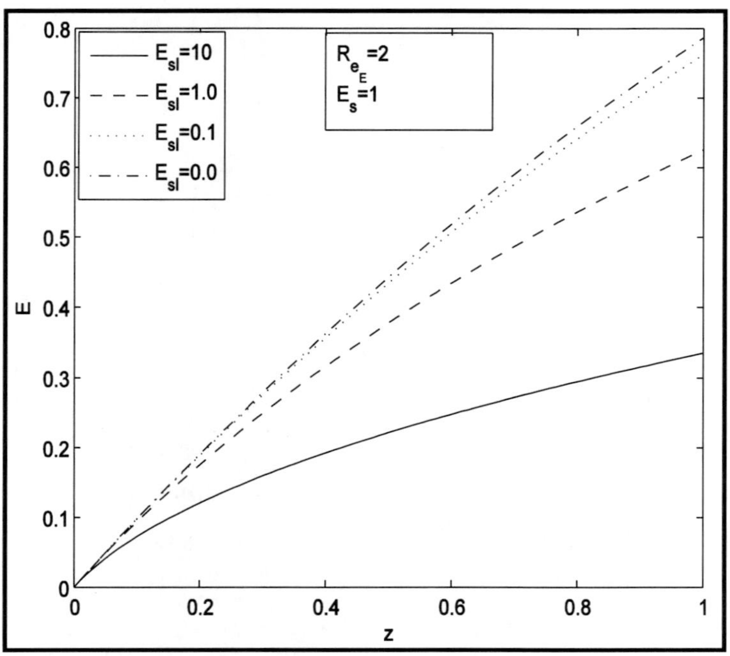

Figure 4. Distribution of electrical field for $Re_E = 2$, $Es = 1$ for various E_{sl} values.

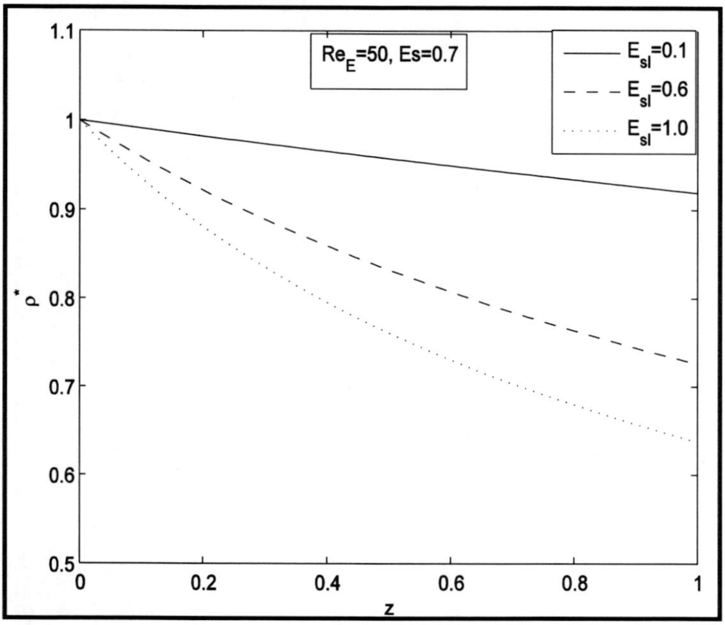

Figure 5. Charge density variation with $Re_E = 50$, $Es = 0.7$ for various E_{sl}.

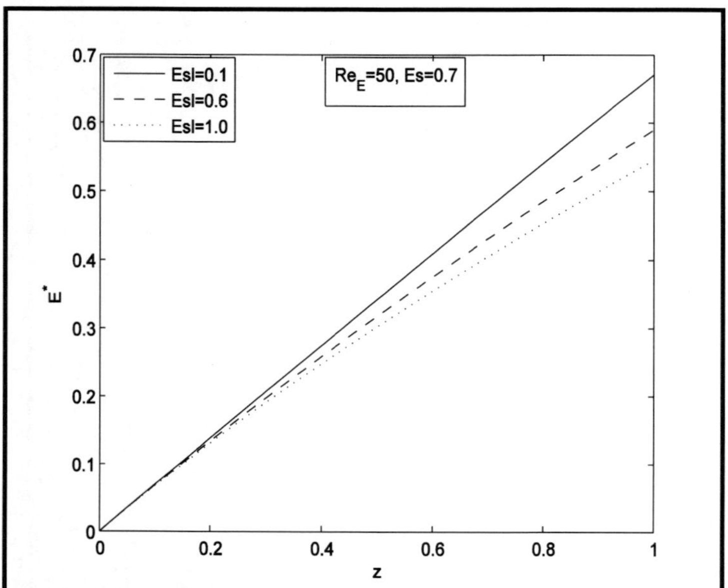

Figure 6. Electrical field profiles with $Re_E = 50$, $Es = 0.7$ for various E_{sl}.

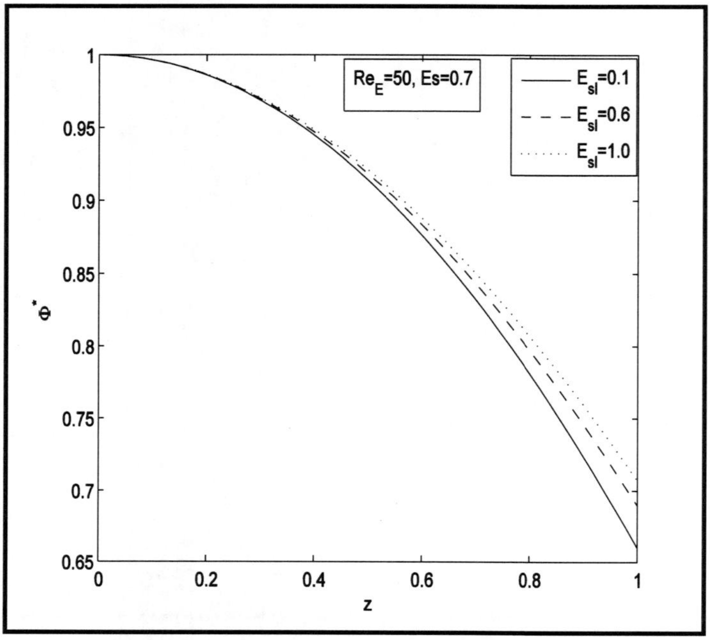

Figure 7. Electric potential variation with $Re_E = 50$, $Es = 0.7$ for various E_{sl}.

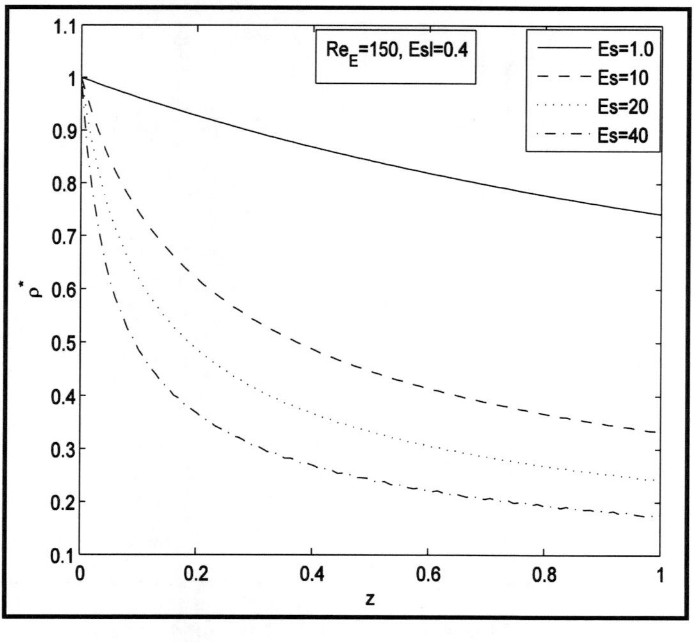

Figure 8. Charge distribution variation with $Re_E = 150$, $E_{sl} = 0.4$ for various E_s.

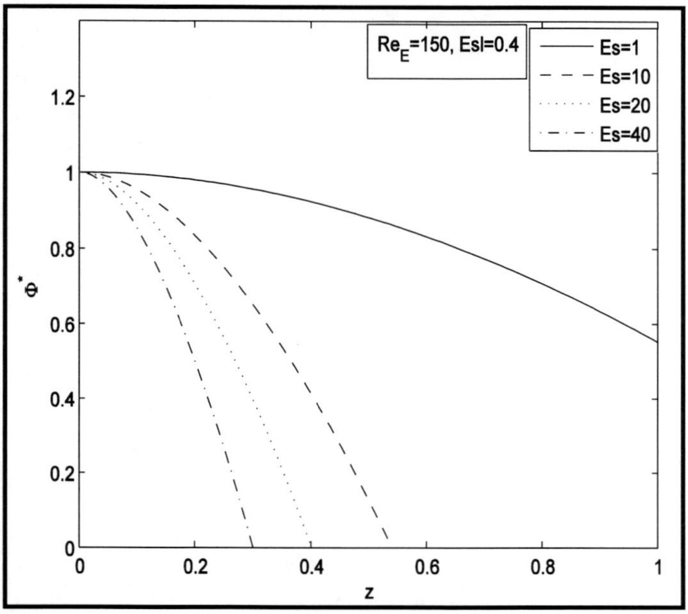

Figure 9. Electrical potential variation with $Re_E = 150$, $E_{sl} = 0.4$ for various E_s.

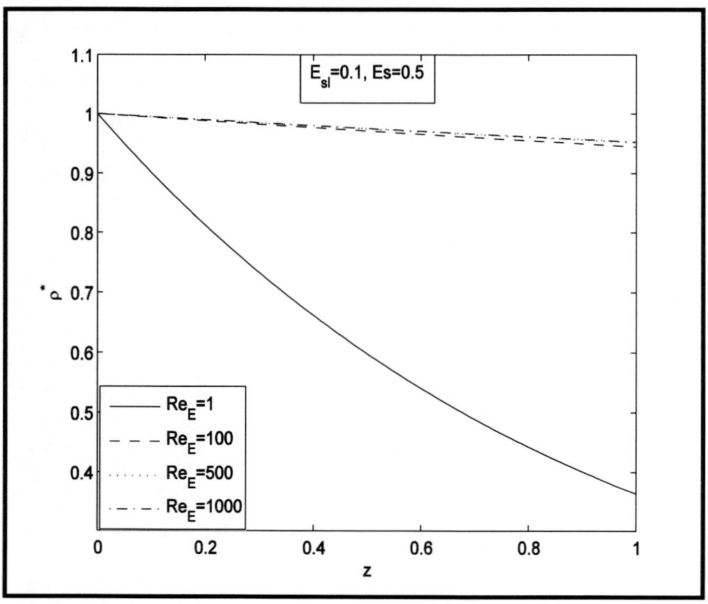

Figure 10. Charge distribution profiles for $E_{sl} = 0.1$, $Es = 0.5$, with various Re_E.

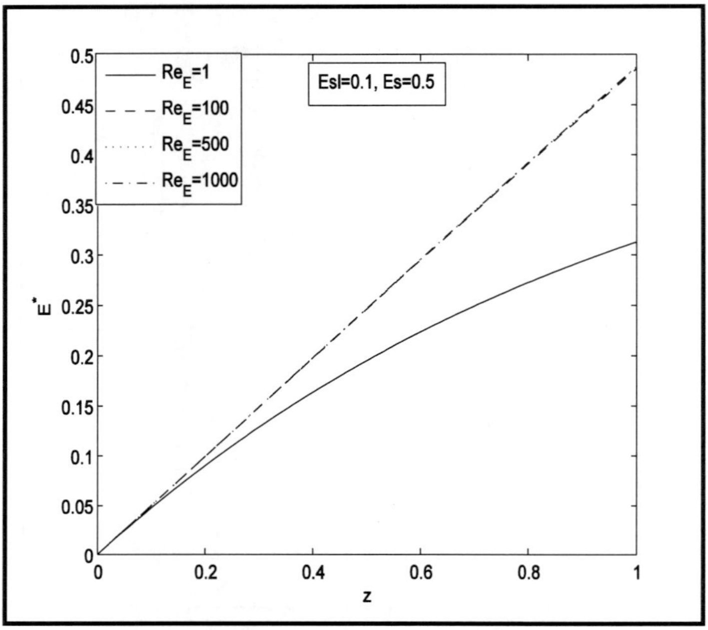

Figure 11. Electrical field profiles with $E_{sl} = 0.1$, $Es = 0.5$, for various Re_E.

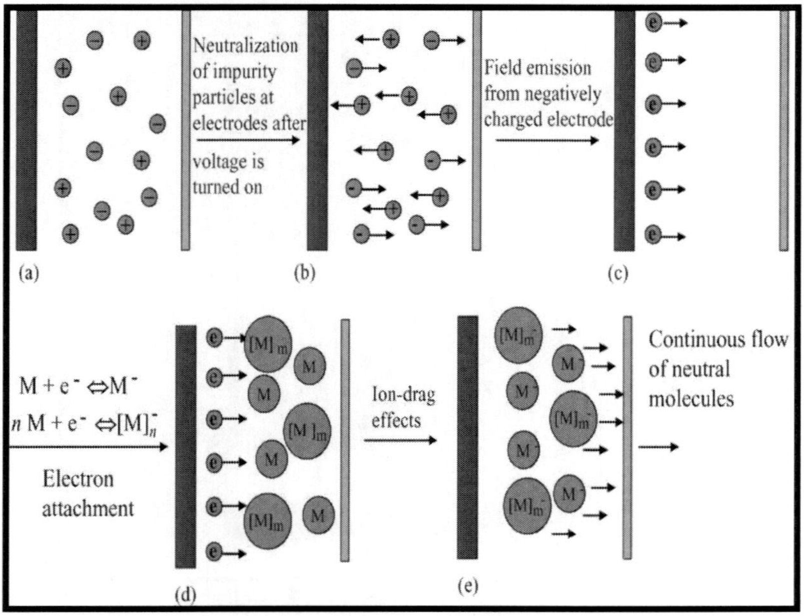

Figure 12. Transient and more complex EHD model for ion drag.

Finally Figures 10, 11 show the influence of electrical Reynolds number on the charge and electrical field distributions. Increasing Re_E strongly elevates the charge distribution magnitudes. For $Re_E = 1$ (lowest value studied) the charge density values are therefore minimized. This case corresponds to the situation where the inertial force and the electrical force in the ion flow are equal. For $Re_E > 1$, inertial force dominates the electrical force in the fluid, and this serves to accentuate charge density distribution values. For $Re_E < 1$, which is not considered in Figure 21) the ion flows will be decelerated due to a reduction in inertial (momentum) force and the charge density will also be reduced markedly, as indicated by Yanada et al. [28] and also by Stuetzer [29]. We note that at extremely high Re_E (500 or above), profiles attain a limit i.e., there is a limiting value of charge density achievable in real flows. Electrical field is also observed to be strongly enhanced with increasing electrical Reynolds number in Figure 11. However we observe that for $Re_E = 100, 500, 1000$, there is no change in electrical field. Clearly therefore an upper limit exists to the maximum electrical field which can be generated in the flow and this limiting scenario takes a linear profile, compared with the distinctly monotonic profile at very low electrical Reynolds number ($Re_E = 1$). Overall the **SCM** method has generated very accurate and physically correct solutions to both electrohydrodynamic flow problems considered. Solutions in all cases have been found to be both stable and convergent. Generalizations of this flow model have also been explored for time-dependence and impurities – see Figure 12 [30].

3. MAGNETOHYDRODYNAMIC THRUSTERS

We now turn our attention to magnetic field propulsion systems. The magneto-plasma-dynamic (MPD) thruster is currently the most powerful form of electromagnetic propulsion. The MPD's ability to efficiently convert megawatts of electric power into thrust makes this technology a prime candidate for economical delivery of lunar and Mars cargo, outer planet rendezvous, and sample return, and for enabling other bold new ventures in deep space robotic and piloted planetary exploration. With its high exhaust velocities, MPD propulsion offers distinct advantages over conventional types of propulsion for each of these mission applications. MPDs expel plasma to create propulsion. MPDs can process more power and create more thrust than any other type of electric propulsion currently available, while maintaining the high exhaust velocities associated with ion propulsion. The general MPD thruster has two metal electrodes: a central rod-shaped cathode, and a cylindrical anode that surrounds the cathode. Just as in an arc welder, a high-current electric arc is struck between the anode and cathode. As the cathode heats up, it emits electrons, which collide with and ionize a propellant gas to create plasma.

A magnetic field is created by the electric current returning to the power supply through the cathode, just like the magnetic field that is created when electrical current travels through a wire. This self-induced magnetic field interacts with the electric current flowing from the anode to the cathode (through the plasma) to produce an electromagnetic (Lorentz) force (cross product of current density and magnetic field) that pushes the plasma out of the engine, creating thrust. An external magnet coil may also be used to provide additional magnetic fields to help stabilize and accelerate the plasma discharge. An excellent perspective of historical developments is given elsewhere [31]. The leading research centres currently are Caltech's NASA Jet Propulsion Laboratory and Princeton University, where lithium-fed MPD thrusters have been established. Hydrogen-based test facilities have been established at the NASA Glenn Research in California. Lithium-fed MPD thrusters developed in Russia have operated at power levels of 100 kilowatts, with efficiencies of up to 45 percent and plasma exhaust velocities approaching 50,000 meters per second (over 100,000 miles per hour). Some examples of current systems under development are shown in Figures 13-15.

A promising concept, developed at Florida is the vapour core MHD superconducting swirl disk reactor, illustrated in Figure 15. In this design, a gas core reactor is coupled to a disk MHD unit with superconducting magnets. This mechanism provides a robust methodology for high performance topping cycle in MHD vapour core reactor spacecraft which could yield 70% efficiency or even greater. The proposed design would be a high-fuel burn-up system with online extraction of fission products. This system also has excellent characteristics for environmental and long-term economic viability and would achieve a completely closed fuel cycle.

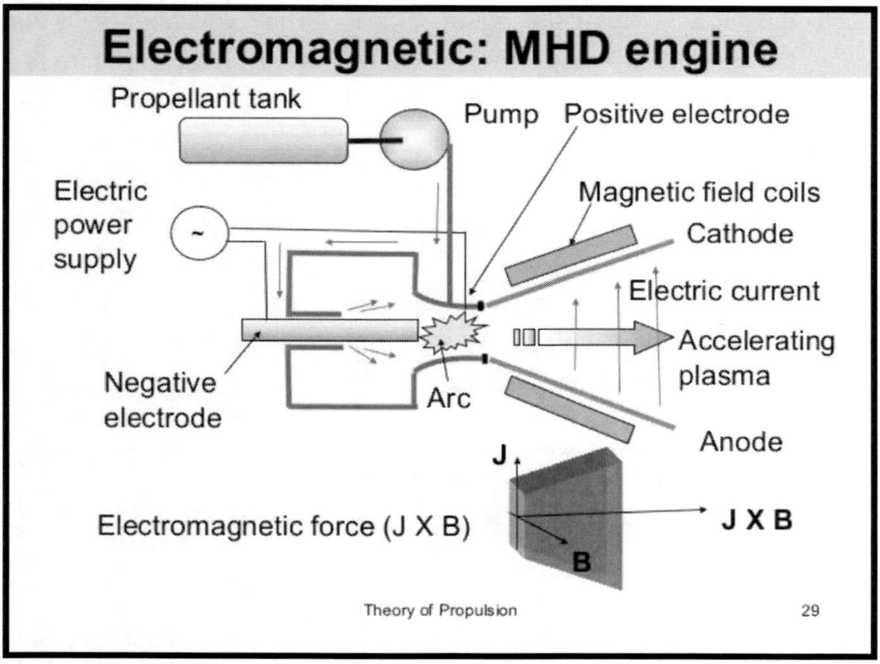

Figure 13. Caltech/NASA generic MPD thruster engine [courtesy-NASA Glenn].

The unspent fuel remaining from the reactor loop shut down during plant decommissioning does not posed any long term storage problems. In space travel, it may easily in fact be ejected. The swirl disk component of this system involves magnetic-viscous flow and entropy generation. This system is also popular in rotating MHD energy generators as shown in Figure 16. In both systems *high temperature electrically-conducting gas (plasma)* enters the system under strong magnetic field. The plasma exits the channel through a magnetic field although the magnet conductivity might be small. In whichever case, the magnet must be constructed from a superconductor consisting of high cost of construction alongside topping units which include the combustor, nozzle accelerator, diffuser etc. Swirling disk flows are known as *Von Karman* flows. They constitute a rare case for reducing the full Navier-Stokes viscous flow partial differential equations to coupled, ordinary differential equations, massively simplifying numerical solutions. In this EMP system we employ DTM-Padé computational simulation to study the entropy generation (via the second law of thermodynamics) in magneto-gas dynamic flow from a stretching rotating disk with heat transfer. The *second law of thermodynamics* is employed by engineers to obtain the optimal design of thermal systems via minimizing the irreversibility and entropy generation, which can improve the efficiency of industrial systems [32-34]. Entropy generation involves thermodynamic irreversibilities, such as characteristics of convective heat transfer, heat transfer across finite temperature gradients, viscous dissipation effects and magnetic field effects which arise in modern aerospace heat transfer processes [35].

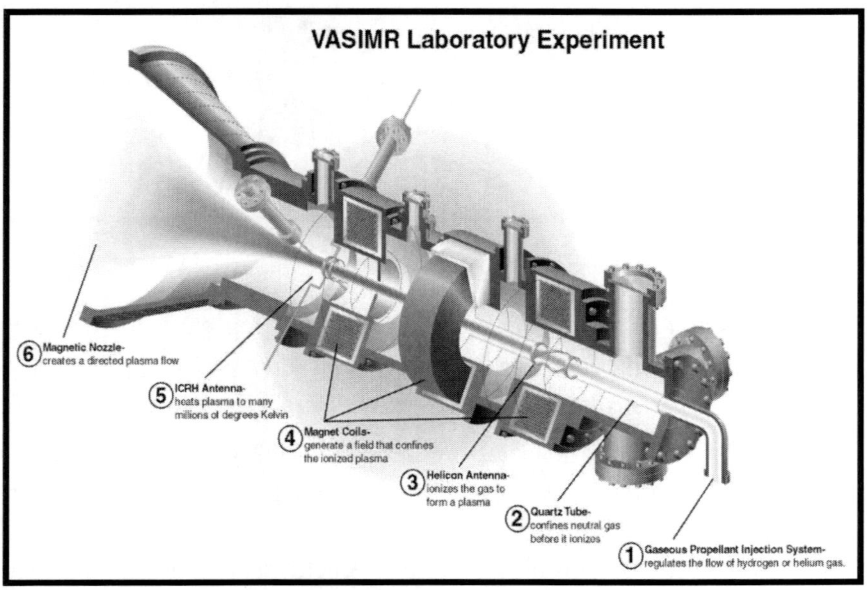

Figure 14. ESA/Princeton VASIMR MPD system.

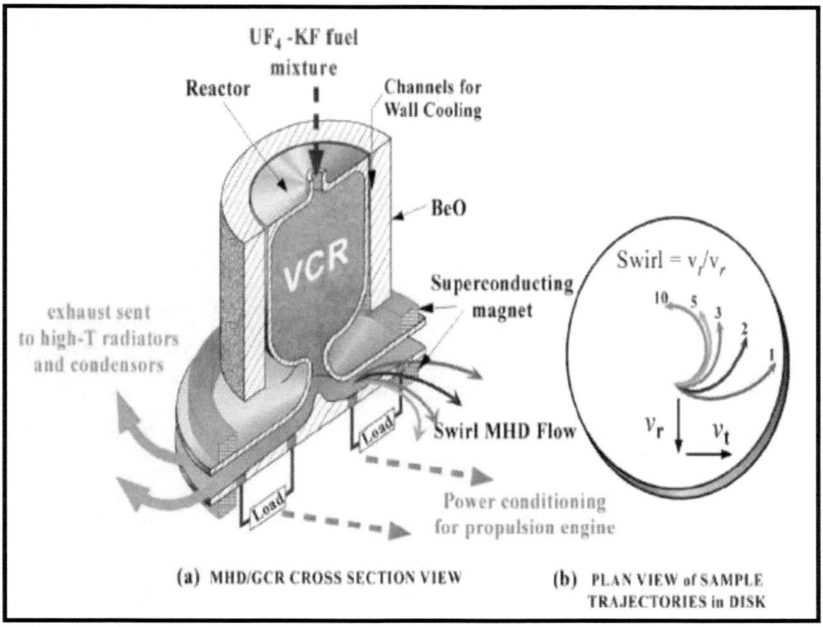

Figure 15. Swirling magnetohydrodynamic entropy nuclear propulsion system.

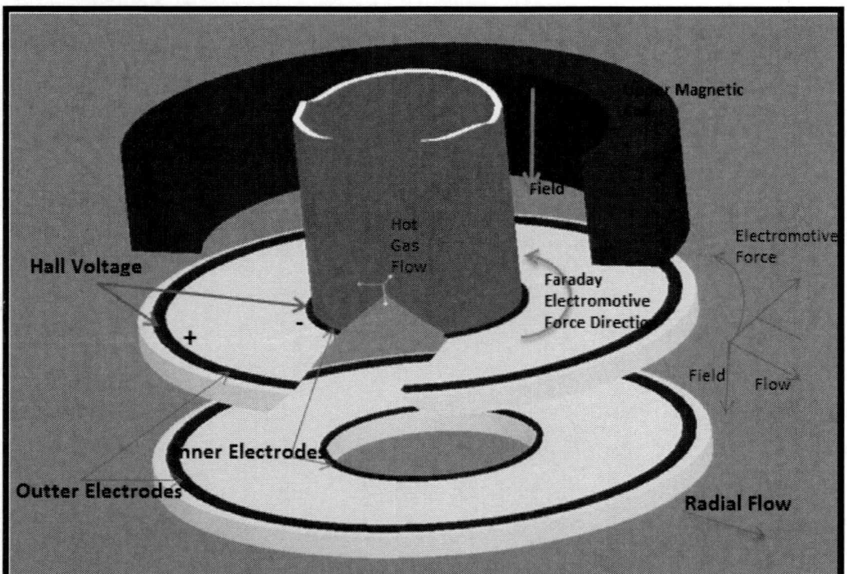

Figure 16. Rotating MHD disk energy generator.

The fluid flow and heat transfer processes are *intrinsically* irreversible, which leads to an increase entropy generation and useful energy destruction [36]. Entropy generation has stimulated significant interest in recent years in aerospace thermal sciences. Erbay et al. [37] used second law analysis to investigate computationally the entropy generation in a channel with a finite volume method and **SIMPLE** algorithm. Al-Odat et al. [38] simulated the effect of magnetic field on the entropy generation due to laminar forced convection over a horizontal flat plate with an implicit finite difference technique. Aïboud and Saouli [39] studied second law thermodynamics and viscoelastic magnetized flow from a stretching surface with double-diffusive convective heat and mass transfer, elaborating in detail the influence of magnetic parameter, Reynolds number and Prandtl number on the entropy generation number. Sahin [40] described the effect of the variable viscosity on the entropy generation in a laminar fluid flow, showing that the entropy generation due to viscous friction becomes dominant for low heat-flux terms. Ibáñez and Cuevas [41] examined entropy generation minimization of a hydromagnetic flow in a micro-channel. Makinde and Bég [42] employed a perturbation expansion technique coupled with a special Hermite-Pade' approximations in the **MAPLE** program, to investigate volumetric entropy generation numbers, irreversibility distribution ratio and the Bejan number evolution in magnetohydrodynamic channel flow in a plasma propulsion duct. Arikogluet al. [43] reported on the effect of slip in entropy generation from rotating disk in hydromagnetic flow with a differential transform method. Entropy generation analysis was applied to modeling and optimization of MHD induction devices by Salas et al. [44]. Rotating magneto-hydrodynamic (MHD) flows exploit magnetic fields which can induce currents in a movable conductive fluid. Liquid metals, plasmas and electrolytes are all important examples of MHD fluids. MHD flows also utilize a Lorentzian drag force which can be used to accelerate/decelerate flows. Many spacecraft applications of rotating hydromagnetic flows including swirling disk flows, flows from revolving cones in liquid metal stirring etc, are documented in the recent monograph by Bég et al. [45]. In order to describe many physical systems with a mathematical model, nonlinear equations are used. Numerical methods are frequently deployed to solve these complex systems of differential equations, which may be multi-degree and often strongly coupled. Although many powerful numerical techniques exist including finite element methods [46], network simulation [47], finite difference methods [48, 49] and CFD codes such as FLUENT [50], computational expense is always a problem and boundary conditions can be difficult to address. Therefore semi-numerical/analytical methods, such as the homotopy perturbation method [51] (**HPM**), differential transform method (**DTM**) [52] and homotopy analysis method (**HAM**) [53] have found increasing popularity amongst fluid dynamics researchers. These techniques can successfully tackle nonlinear differential equations for a variety of boundary conditions, and therefore hold significant promise in propulsion engineering sciences. In this section of the present chapter, we shall review a robust model developed recently for swirling MPD thruster nuclear hybrid engines [54]. As the second law analysis is more reliable than first law analysis for many aerospace and nuclear engineering systems, entropy generation analysis is also included. The effect of various thermophysical parameters, such as *magnetic interaction parameter, rotation parameter, Eckert number, Prandtl number, Brinkman number* and *Reynolds number* on velocity and temperature fields and also on the entropy generation and Bejan number are investigated in detail. An excellent summary of viscous magneto-hydrodynamics is given by Chandrasekhar [55]. In the present model, which is *sub-sonic*, we ignored Alfven wave effects, Hall currents and ionslip phenomena. The 3-dimensional steady magneto-hydrodynamic laminar boundary-layer flow of a Newtonian, electrically-conducting, viscous fluid from a rotating disk in the presence of an externally applied axially-directed uniform magnetic field, (*hydromagnetic Von Karman swirling plasma flow)* can be simulated using the continuity, Navier-Stokes (momentum conservation), generalized Ohm's law and energy conservation equations, and these may be presented, in the presence of Ohmic (Joule) heating and viscous dissipation, in vectorial form, as follows [56]:

$$\nabla \cdot \boldsymbol{u} = 0 \tag{27}$$

$$\rho(\boldsymbol{u} \cdot \nabla)\boldsymbol{u} = -\nabla P + \mu \nabla^2 \boldsymbol{u} + \boldsymbol{J} \times \boldsymbol{B}, \tag{28}$$

$$\boldsymbol{J} = \sigma[\boldsymbol{E} + \boldsymbol{u} \times \boldsymbol{B}] \tag{29}$$

$$\rho c_p (\mathbf{u} \cdot \nabla) T = k \nabla^2 T + \mu \Phi + \frac{J^2}{\sigma}. \tag{30}$$

where $\boldsymbol{u}$ is velocity vector, ρ is the density of the conducting gas, μ is dynamic viscosity of gas, $\boldsymbol{B}$ is external uniform magnetic field, B_e is constant magnetic flux density, B_o is constant magnetic flux density, c_p is specific heat at constant pressure, $\boldsymbol{E}$ is electrical field, $\boldsymbol{J}$ is current density field, k is thermal conductivity, P is pressure, σ is electrical conductivity of gas (plasma), T is temperature, Φ is viscous dissipation function.

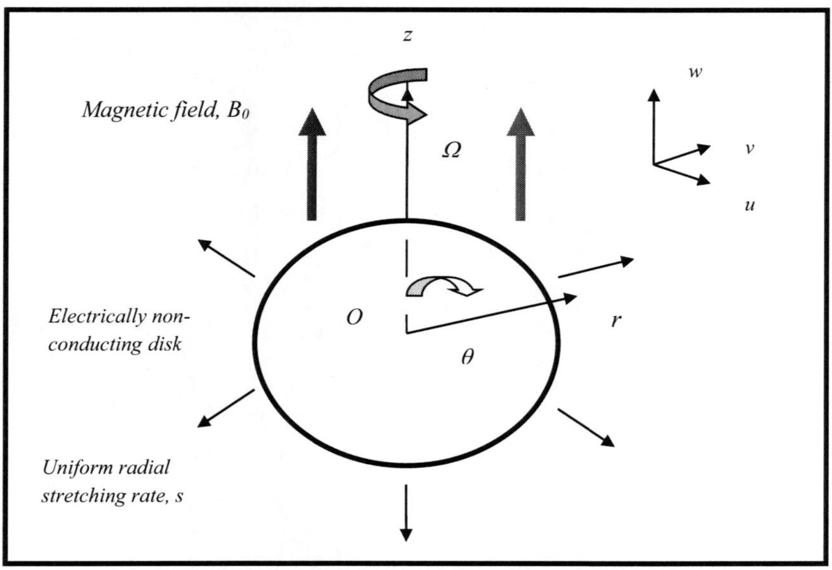

Figure 17. 3-D swirling Von Karman magnetohydrodynamic flow regime and coordinate system.

The disc radius is much larger than the boundary layer thickness, so that edge effects can be neglected. The applied magnetic field is steady and sufficiently weak to neglect the induced magnetic field i.e., a *low magnetic Reynolds number* is assumed. This is a reasonable assumption for the flow of certain working fluids in nuclear-MHD propulsion systems [57] and also in nuclear engineering liquid metals, e.g., liquid sodium [58]. The physical regime is depicted in Figure 17 with reference to non-rotating cylindrical polar coordinates (r, θ, z).

The disk rotates with a constant angular velocity (Ω) and also stretches with a constant rate (s) in the radial direction [59]. The external uniform magnetic field $\boldsymbol{B}$ is applied transverse to the disk plane i.e., in the z-direction and possesses a constant magnetic flux density B_0. Furthermore all thermophysical and fluid properties are assumed to be constant and radial and tangential electrical currents are neglected. The disk itself is electrically non-conducting, so that current density vanishes both at the disk surface and in the fluid regime. The penultimate and final terms on the right hand side of eqn. (30) designate the *viscous heating* and *Joule heating* contributions, respectively, and have been shown to be significant in propulsion applications [60]. The non-dimensional form for the mean flow velocities and temperature distribution are provided by Von Karman's classical transformations [61]. With the aid of a dimensionless normal distance from the wall, $\eta = zR$, and an appropriate Reynolds number incorporating the disk stretching rate, $R = \left(s/\upsilon\right)^{1/2}$ we have:

$$(u, v, w, \mathrm{T}) = \left(rsF(\eta), rsG(\eta), \frac{1}{R}H(\eta),\ \mathrm{T}_\infty + (\mathrm{T}_w - \mathrm{T}_\infty)\theta(\eta) \right), \tag{31}$$

where the dimensionless functions *F, G, H*, and *T* (which denote self-similar radial, azimuthal, axial velocity and temperature, respectively) satisfy the following ordinary differential equations defining the radial, azimuthal, axial momentum conservation and energy conservation in the regime:

$$H' + 2F = 0, \tag{32}$$

$$F'' - F^2 + G^2 - HF' - MF = 0, \tag{33}$$

$$G'' - 2FG - HG' - MG = 0, \tag{34}$$

$$\theta'' - Pr\theta' H + Ec\left(F'^2 + G'^2\right) + EcM\left(F^2 + G^2\right) = 0. \tag{35}$$

The boundary conditions at the stretching disk surface and in the free stream, are prescribed respectively, as:

$$\begin{aligned} &H = F - 1 = G - \omega = \theta - 1 = 0, && \text{at } \eta = 0, \\ &F = G = \theta = 0, && \text{as } \eta \to \infty, \end{aligned} \tag{36}$$

Here $\omega = \Omega/s$ denotes the "rotation strength parameter" expressing the ratio of disk swirl to disk radial stretch. The case of $\omega = 0$ corresponds to *pure stretching* of the disk without rotation. The nonlinear, coupled ODEs defined in eqns. (32)–(35) under boundary conditions (36) constitute a robust two-point boundary value problem, which we solve using a combination of the differential transform method (DTM) and Padé approximants, namely **DTM-Padé** simulation [62].

3.1. Entropy Generation Analysis

The volumetric rate of local entropy generation, for the present hydromagnetic swirling heat transfer problem, in the presence of the axial symmetry, can be presented as follows:

$$\dot{S}_{gen}^{///} = \frac{k}{T_o^{\,2}}[\nabla T]^2 + \frac{\mu}{T_o}\Phi + \frac{1}{T_o}[(\boldsymbol{J} - \boldsymbol{Q}\boldsymbol{V}).(\boldsymbol{E} + \boldsymbol{V} \times \boldsymbol{B})] \tag{37}$$

Here equation (3.29) also applies and the following definitions:

$$[\nabla \mathrm{T}]^2 = \left[\left(\frac{\partial \mathrm{T}}{\partial r}\right)^2 + \left(\frac{1}{r}\frac{\partial \mathrm{T}}{\partial \theta}\right)^2 + \left(\frac{\partial \mathrm{T}}{\partial z}\right)^2\right], \tag{38}$$

$$\Phi = 2\left[\left(\frac{\partial u}{\partial r}\right)^2 + \frac{1}{r^2}\left(\frac{\partial v}{\partial \theta} + u\right)^2 + \left(\frac{\partial w}{\partial z}\right)^2\right] + \left(\frac{\partial v}{\partial z} + \frac{1}{r}\frac{\partial w}{\partial \theta}\right)^2$$
$$+\left(\frac{\partial w}{\partial r} + \frac{\partial u}{\partial z}\right)^2 + \left[\frac{1}{r}\frac{\partial u}{\partial \theta} + r\frac{\partial}{\partial r}\left(\frac{v}{r}\right)\right]^2, \tag{39}$$

It is assumed that electric force per unit charge is negligible compared with $\boldsymbol{V} \times \boldsymbol{B}$ in Eqs. (37) and furthermore that electric current is much greater than $\boldsymbol{QV}$. With these assumptions, Eq. (37) can be shown to take the form:

$$\dot{S}'''_{gen} = \frac{k}{T_0^2}\left(\frac{\partial \mathrm{T}}{\partial z}\right)^2 + \frac{\mu}{T_0}\left\{2\left[\left(\frac{\partial u}{\partial r}\right)^2 + \frac{1}{r^2}u^2 + \left(\frac{\partial w}{\partial z}\right)^2\right] + \left(\frac{\partial v}{\partial z}\right)^2 + \left(\frac{\partial u}{\partial z}\right)^2 + \left[r\frac{\partial}{\partial r}\left(\frac{v}{r}\right)\right]^2\right\} + \frac{\sigma \mathrm{B}_0^2}{\mathrm{T}_0}\left(u^2 + v^2\right). \tag{40}$$

$\boldsymbol{Q}$ denotes the electrical charge density, $\dot{S}'''_{gen}$ is the *volumetric* rate of local entropy generation. The right hand side of the above entropy generation equation consists of three parts. The first term is the *local entropy generation* due to heat transfer irreversibility. The second larger part refers to the fluid friction irreversibility. The final part denotes magnetic field effects. The dimensionless form of the entropy generation rate, termed *entropy generation number*, defines the ratio between the *actual entropy generation rate* $\left(\dot{S}'''_{gen}\right)$ and the *characteristic entropy generation rate* $\left(\dot{S}'''_0\right)$. By using the Von Karman transformation parameters, as defined in Eqn. (31), the entropy generation number $\left(N_G\right)$ becomes, for the present system:

$$\mathrm{N}_G = \alpha\theta'(\eta)^2 + \frac{4\mathrm{B}r}{Re}F(\eta)^2 + \frac{2Ec}{Re}H'(\eta)^2 + Br\left(F'(\eta)^2 + G'(\eta)^2\right) + Br \times M\left(F(\eta)^2 + G(\eta)^2\right), \tag{41}$$

Inspection of Eq. (41), indicates that with increasing value of α, the effect of *heat transfer irreversibility* increases. As the Brinkman number and Eckert number increase or the Reynolds number decreases, the entropy generation due to fluid friction irreversibility *increases*. Irreversibility mechanism domination is an important aspect of entropy generation analysis since the entropy generation number does not provide any information regarding this. The *Bejan number* and the *irreversibility distribution ratio* are introduced to overcome this shortcoming. The Bejan number $\left(Be\right)$ embodies the ratio of *entropy generation which is caused through heat transfer* to the *total entropy generation*. The irreversibility distribution ratio $\left(\phi\right)$ expresses the ratio between *entropy generation due to fluid friction* and *joule dissipation in heat transfer*. These two parameters, for the present swirling magnetohydrodynamic thermal problem can be expressed as:

$$Be = \frac{\alpha\theta'(\eta)^2}{\alpha\theta'(\eta)^2 + \frac{4\mathrm{B}r}{Re}F(\eta)^2 + \frac{2Ec}{Re}H'(\eta)^2 + Br\left(F'(\eta)^2 + G'(\eta)^2\right) + Br \times M\left(F(\eta)^2 + G(\eta)^2\right)}, \tag{42}$$

$$\phi = \frac{\frac{4\mathrm{B}r}{Re}F(\eta)^2 + \frac{2Ec}{Re}H'(\eta)^2 + Br\left(F'(\eta)^2 + G'(\eta)^2\right) + Br \times M\left(F(\eta)^2 + G(\eta)^2\right)}{\alpha\theta'(\eta)^2}. \tag{43}$$

The behavior of the Bejan number and irreversibility distribution ratio are almost the same. Thus we only examine the Bejan number effects in this article. The Bejan number falls in the range, $0< Be < 1$. When $Be = 0$, *fluid friction irreversibility* dominates. For $Be = 0.5$, heat transfer and fluid friction irreversibility effects are the same. For $Be = 1$, the irreversibility mechanism is dominated by *heat transfer* effects.

3.2. Differential Transform Method (DTM) Simulation

DTM is employed in conjunction with Padé approximants to solve the governing equations (32) to (35), under boundary conditions (36). DTM was originally developed for electrical circuit modeling by Zhou [63]. The principal attraction of this technique is that it can be applied directly to nonlinear differential equations without requiring discretization and linearization, a strong advantage over purely numerical methods such as finite elements. DTM also does not need require any perturbation parameters and this offers a great advantage over asymptotic and perturbation expansion techniques. This method has been successfully implemented in a diverse array of engineering science problems in recent years including stagnation-point hydromagnetic flow in permeable materials [64], vibration analysis [65], mixed convection flows [66], liquid film magnetic flows [67] and heat and mass transfer in Von Karman swirling hydromagnetic flow [68]. Further applications include radiative heat transfer in non-Newtonian micropolar flow [69], nanofluid convection [70], chemically-reacting mixed convection flows [71] and magneto-hemodynamics [72]. In order to solve highly nonlinear differential equation systems, arising frequently in propulsion fluid dynamics, it is pertinent to combine DTM and Padé approximations and use **DTM-Padé** simulation. This method effectively increases the convergence of DTM. To provide readers with an appreciation of this methodology, we include here a short description of the fundamentals of DTM. The differential transform of the function $u(\eta)$ in one variable, is given by:

$$U(k) = \frac{1}{k!}\left[\frac{d^k u(\eta)}{d\eta^k}\right]_{\eta=\eta_0}, \tag{44}$$

where $U(k)$ is the transformed function and $u(\eta)$ is the original function. The reverse transformation of $U(k)$ is defined as follows:

$$u(\eta) = \sum_{k=0}^{\infty}(\eta - \eta_0)^k U(k), \tag{45}$$

$$u(\eta) \cong \sum_{k=0}^{N}(\eta - \eta_0)^k U(k), \tag{46}$$

In practical engineering applications, a *finite series* is used to evaluate the function $u(\eta)$. Therefore, Eq. (45) can be written in the form of Eq. (46). This implies that $\sum_{k=N+1}^{\infty}(\eta - \eta_0)^k U(k)$, is omitted, since this is negligible in comparison with the other terms.

3.3. The Padé Approximants

Numerous mathematical techniques exist to increase the *convergence radius* of a given series. Assuming a function $u(\eta)$ which is expressed by a power series $\sum_{i=0}^{\infty} a_i \eta^i$, so that:

$$u(\eta) = \sum_{k=0}^{\infty} a_k \eta^k, \tag{47}$$

The Padé approximant is a *rational fraction* and the notation for such a Padé approximant is [73]:

$$[L, M] = \frac{P_L(\eta)}{Q_M(\eta)}, \tag{48}$$

where $P_L(\eta)$ is a polynomial with highest degree L and $Q_M(\eta)$ is a polynomial with highest degree M. We have:

$$u(\eta) = a_0 + a_1\eta + a_2\eta^2 + a_3\eta^3 + a_4\eta^4 + \cdots \tag{49}$$

$$P_L(\eta) = p_0 + p_1\eta + p_2\eta^2 + p_3\eta^3 + \cdots + p_L\eta^L, \tag{50}$$

$$Q_M(\eta) = q_0 + q_1\eta + q_2\eta^2 + q_3\eta^3 + \cdots + q_M\eta^M, \tag{51}$$

It should be emphasized that in Eq. (23) there are $L+1$ numerator coefficients and $M+1$ denominator coefficients. Hence we can clearly multiply the numerator and denominator by a constant and leave $[L, M]$ unchanged. In light of this, we exert the *normalization* condition:

$$Q_M(0) = 1, \tag{52}$$

By applying the above condition, it can be concluded that there are $L+1$ independent numerator coefficients and M independent denominator coefficients, and this leads to $L+M+1$ unknown coefficients in all. This number indicates that normally $[L/M]$ ought to fit the power series in Eq. (47) through the orders $1, \eta, \eta^2, \ldots, \eta^{L+M}$. As elaborated by

Baker and Graves-Morris [**74**] the $[L/M]$ approximant is *uniquely* determined. Thus, according to the above equations, we have:

$$\sum_{k=0}^{\infty} a_k \eta^k = \frac{p_0 + p_1\eta + p_2\eta^2 + \cdots + p_L\eta^L}{q_0 + q_1\eta + q_2\eta^2 + \cdots + q_M\eta^M} + O\left(\eta^{L+M+1}\right), \tag{53}$$

By cross-multiplying the above equation, we obtain:

$$\left(a_0 + a_1\eta + a_2\eta^2 + a_3\eta^3 + \cdots\right)\left(1 + q_1\eta + q_2\eta^2 + \cdots + q_M\eta^M\right) = p_0 + p_1\eta + p_2\eta^2 + \cdots + p_L\eta^L + O\left(\eta^{L+M+1}\right). \tag{54}$$

From Eq. (54), the following system of equations are readily produced:

$$\begin{cases} a_0 = p_0, \\ a_1 + a_0 q_1 = p_1, \\ a_2 + a_1 q_1 + a_0 q_2 = p_2, \\ \vdots \\ a_L + a_{L-1} q_1 + L + a_0 q_L = p_L, \end{cases} \tag{55}$$

$$\begin{cases} a_{L+1} + a_L q_1 + \cdots + a_{L-M+1} q_M = 0, \\ a_{L+2} + a_{L+1} q_1 + \cdots + a_{L-M+2} q_M = 0, \\ \vdots \\ a_{L+M} + a_{L+M-1} q_1 + \cdots + a_L q_M = 0, \end{cases} \tag{56}$$

where $q_j = 0$ for $j > M$ and $a_n = 0$ for $n < 0$. We can solve the Eqs. (30) and (31) directly, if they are *non-singular:*

$$[L/M]=\frac{\begin{vmatrix} a_{L-M+1} & a_{L-M+2} & \cdots & a_{L+1} \\ \vdots & \vdots & \ddots & \vdots \\ a_L & a_{L+1} & \cdots & a_{L+M} \\ \sum_{j=M}^{L} a_{j-M}x^j & \sum_{j=M-1}^{L} a_{j-M+1}x^j & \cdots & \sum_{j=0}^{L} a_j x^j \end{vmatrix}}{\begin{vmatrix} a_{L-M+1} & a_{L-M+2} & \cdots & a_{L+1} \\ \vdots & \vdots & \ddots & \vdots \\ a_L & a_{L+1} & \cdots & a_{L+M} \\ x^M & x^{M-1} & \cdots & 1 \end{vmatrix}}. \tag{57}$$

If the lower index on a sum goes beyond the upper, the sum is replaced by zero. Alternate forms are:

$$[L/M]=\sum_{j=0}^{L-M} a_{j-M}\eta^j+\eta_{L-M+1}w_{L/M}^T W_{L/M}^{-1} w_{L/M}=\sum_{j=0}^{L+n} a_j\eta^j+\eta^{L+n+1}w_{(L+M)/M}^T W_{L/M}^{-1} w_{(L+n)/M}, \tag{58}$$

$$W_{L/M}=\begin{bmatrix} a_{L-M+1}-xa_{L-M+2} & \cdots & a_L-xa_{L+1} \\ \vdots & \ddots & \vdots \\ a_L-xa_{L+1} & \cdots & a_{L+M+1}-xa_{L+M} \end{bmatrix}, \tag{59}$$

$$w_{L/M}=\begin{bmatrix} a_{L-M+1} \\ a_{L-M+2} \\ \vdots \\ a_L \end{bmatrix}. \tag{60}$$

The construction of $[L/M]$ approximants includes only algebraic operations [75]. The major difficulty in applying this technique is to direct the choice of numerator and denominator degrees, in order to obtain the *best approximant*. A good criterion which is used in this regard, depending on the shape of the solution, is the judicious selection of $[L/M]$ approximants such that $L = M$. The approximants are generated numerically using ***MATHEMATICA*** software.

3.4. Semi-Numerical Solutions with DTM-Padé Simulation

Let us now consider the swirling magnetohydrodynamic heat transfer regime under investigation. Taking differential transforms of the dimensionless ordinary differential Eqns. (3.32) to (3.35) leads to:

$$(k+1)\,H(k+1)+2\,F(k)=0, \tag{61}$$

$$(k+1)(k+2)F(k+2)-\sum_{r=0}^{k}\left[(k+1-r)H(r)F(k+1-r)+F(r)F(k-r)-G(r)G(k-r)\right]$$
$$-M\,F(k)=0, \tag{62}$$

$$(k+1)(k+2)G(k+2)-\sum_{r=0}^{k}\left[(k+1-r)H(r)G(k+1-r)+2F(r)G(k-r)\right]-M\,G(k)=0, \tag{63}$$

$$(k+1)(k+2)\theta(k+2)$$
$$-\sum_{r=0}^{k}\begin{bmatrix}\Pr(k+1-r)H(r)\theta(k+1-r)\\ +EcM\left(F(r)F(k-r)+G(r)G(k-r)\right)\\ +Ec\left((k-r+1)(r+1)F(r+1)F(k-r+1)+(k-r+1)(r+1)G(r+1)G(k-r+1)\right)\end{bmatrix}=0, \tag{64}$$

where $H(k)$, $F(k)$, $G(k)$ and $\theta(k)$ are the differential transforms of the functions $H(\eta)$, $F(\eta)$, $G(\eta)$ and $\Theta(\eta)$ and are given by:

$$H(\eta)=\sum_{k=0}^{\infty}H(k)\eta^{k}, \tag{65}$$

$$F(\eta)=\sum_{k=0}^{\infty}F(k)\eta^{k}, \tag{66}$$

$$G(\eta)=\sum_{k=0}^{\infty}G(k)\eta^{k}, \tag{67}$$

$$\theta(\eta)=\sum_{k=0}^{\infty}\theta(k)\eta^{k}. \tag{68}$$

The transformed boundary conditions at the disk surface and in the free stream, now become:

$$\begin{aligned}
&H(0)=0,\\
&F(0)=1, \quad F(1)=\alpha,\\
&G(0)=\omega, \quad G(1)=\beta,\\
&\theta(0)=1, \quad \theta(1)=\gamma,
\end{aligned} \tag{69}$$

where α, β and γ are constants. In addition, by making appropriate substitutions and implementing a *recursive* method it is straightforward to evaluate $H(k)$, $F(k)$, $G(k)$ and $\theta(k)$. Therefore, substituting all $H(k)$; $F(k)$, $G(k)$ and $\theta(k)$ into the relevant equations, the following series solutions emerge:

$$H(\eta)=-2\eta-\alpha\eta^{2}+\frac{1}{3}\eta^{3}\left(-1-M+\omega^{2}\right)+\frac{1}{12}\eta^{4}\left(-M\alpha+2\beta\omega\right)+\cdots, \tag{70}$$

$$F(\eta)=1+\alpha\eta+\frac{1}{2}\eta^2\left(1+M-\omega^2\right)+\frac{1}{6}\eta^3\left(M\alpha-2\beta\omega\right)$$
$$+\frac{1}{12}\eta^4\left(-1-M-\beta^2+\omega^2-\omega\left(2\omega+M\omega\right)+\frac{1}{2}M\left(1+M-\omega^2\right)\right)+\cdots, \tag{71}$$

$$G(\eta)=\beta\eta+\omega+\frac{1}{2}\eta^2\left(2\omega+M\omega\right)+\frac{1}{6}\eta^3\left(M\beta+2\alpha\omega\right)$$
$$+\frac{1}{12}\eta^4\left(\alpha\beta+2\omega+M\omega-2\left(2\omega+M\omega\right)+\frac{1}{2}M\left(2\omega+M\omega\right)+\omega\left(1+M-\omega^2\right)\right)+\cdots, \tag{72}$$

$$\theta(\eta)=1+\gamma\eta+\frac{1}{2}\eta^2\left(-Ec\left(\alpha^2+\beta^2\right)-EcM\left(1+\omega^2\right)\right)$$
$$+\frac{1}{6}\eta^3\left(-2\Pr\gamma-2EcM\left(\alpha+\beta\omega\right)-2Ec\left(\beta\left(2\omega+M\omega\right)+\alpha\left(1+M-\omega^2\right)\right)\right)$$
$$+\frac{1}{12}\eta^4\left(\begin{array}{l}-EcM\left(\alpha^2+\beta^2\right)-\Pr\alpha\gamma-2Ec\left(\frac{1}{2}\beta\left(M\beta+2\alpha\omega\right)+\frac{1}{2}\alpha\left(M\alpha-2\beta\omega\right)\right)\\ -2EcM\left(\frac{1}{2}\omega\left(2\omega+M\omega\right)+\frac{1}{2}\left(1+M-\omega^2\right)\right)\\ -Ec\left(\left(2\omega+M\omega\right)^2+\left(1+M-\omega^2\right)^2\right)-2\Pr\left(-Ec\left(\alpha^2+\beta^2\right)-EcM\left(1+\omega^2\right)\right)\end{array}\right)+\cdots. \tag{73}$$

The Padé approximant is applied to *enlarge* the *convergence radius* of the truncated series solution. In this way the polynomial approximation converts into a ratio of two polynomials. Without using the Padé approximant, the DTM solution cannot satisfy boundary conditions at infinity i.e., a marked divergence in solutions results. Therefore it is important to combine the DTM with Padé approximant series solution to provide an effective tool for resolving infinite boundary value problems. Applying the Padé approximation and using the boundary conditions at $\eta=\infty$, yields α, β and γ. The number of required terms is characterized by the convergence of the numerical values to the desired accuracy. Stability, consistency and convergence- three vital requirements for robust numerical algorithms- are successfully achieved with DTM-Padé simulation.

Figure 18. Radial velocity contours with $M = \omega = 1,\ Pr = Ec = 2,\ \upsilon = 9 \times 10^{-6}$ and $s = 10$.

Figure 19. Azimuthal velocity contours with $M = \omega = 1,\ Pr = Ec = 2,\ \upsilon = 9 \times 10^{-6}$ and $s = 10$.

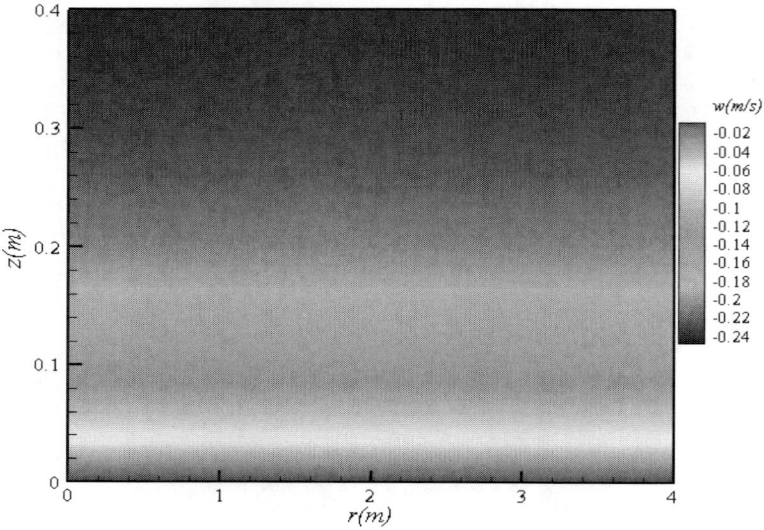

Figure 20. Axial velocity contours with $M = \omega = 1$, $Pr = Ec = 2$, $\upsilon = 9 \times 10^{-6}$ and $s = 10$.

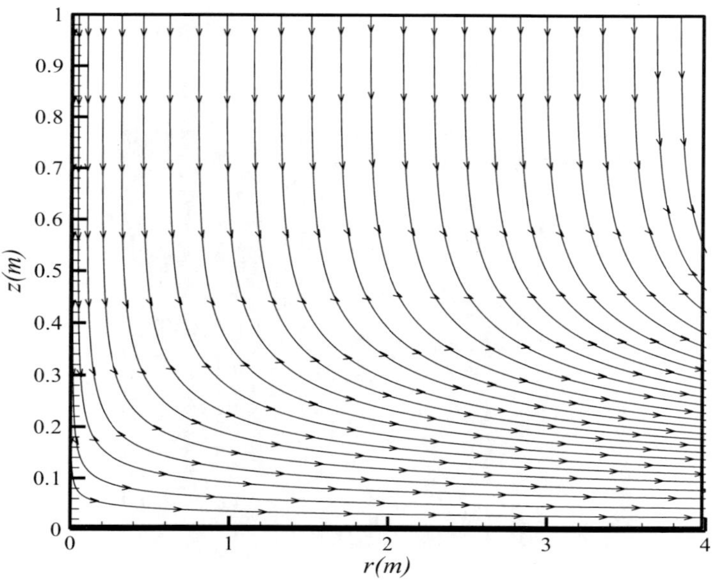

Figure 21. Vector variables of u and w with $M = \omega = 1$, $Pr = Ec = 2$, $\upsilon = 9 \times 10^{-6}$ and $s = 10$.

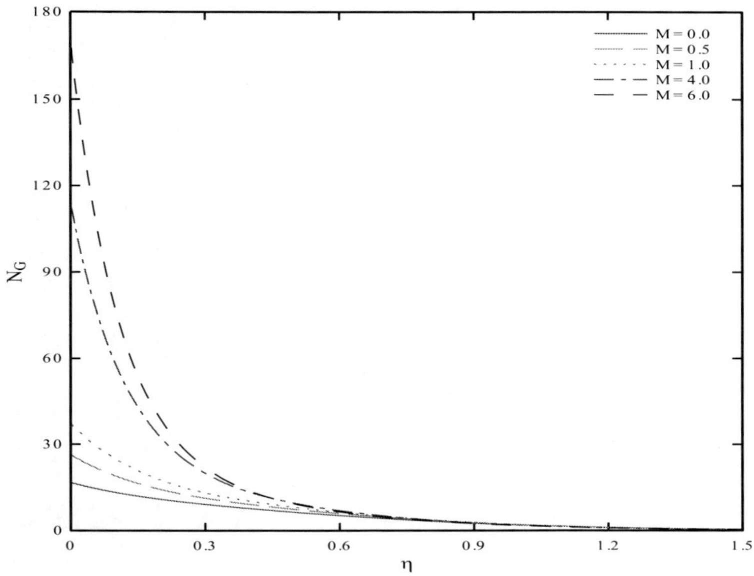

Figure 22. The variation of entropy generation number, N_G with magnetic interaction parameter for $Ec = Pr = 2,\ Br = 4,\ Re = 8$ and $\omega = 1$.

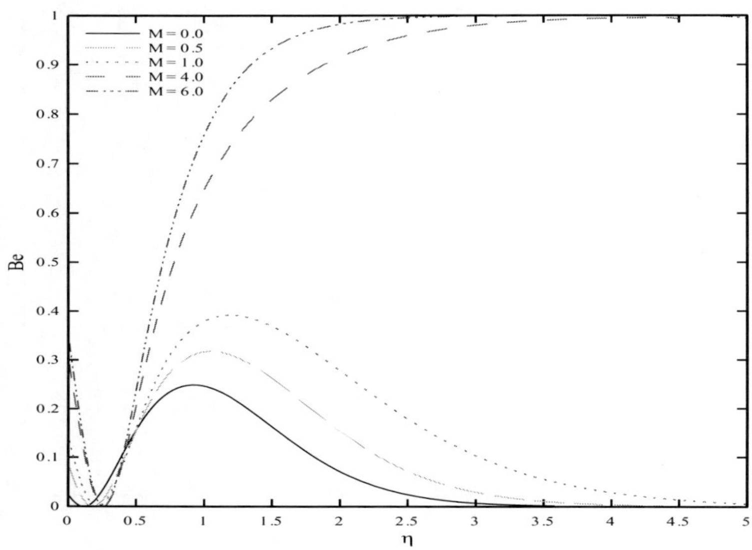

Figure 23. The variation of Bejan number, Be with magnetic interaction parameter for $Ec = Pr = 2,\ Br = 4,\ Re = 8$ and $\omega = 1$.

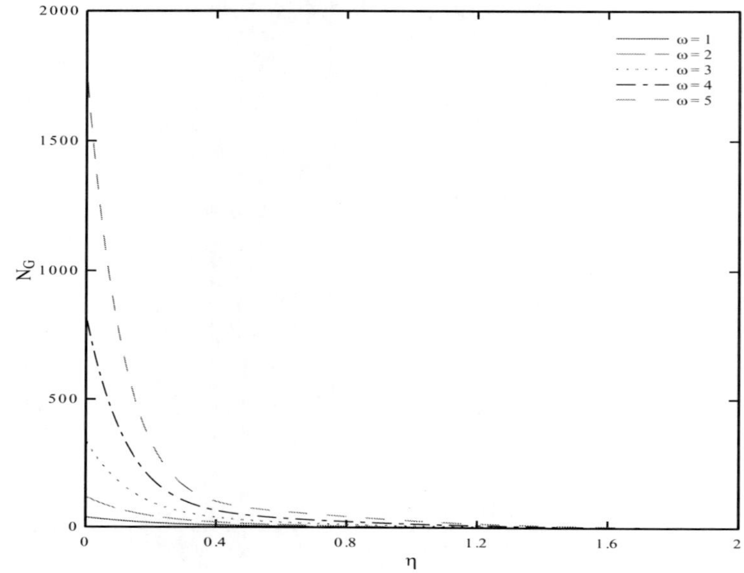

Figure 24. The variation of entropy generation number, N_G with rotation strength parameter for $Ec = Pr = 1,\ M = 2,\ Br = 3$ and $Re = 6.$

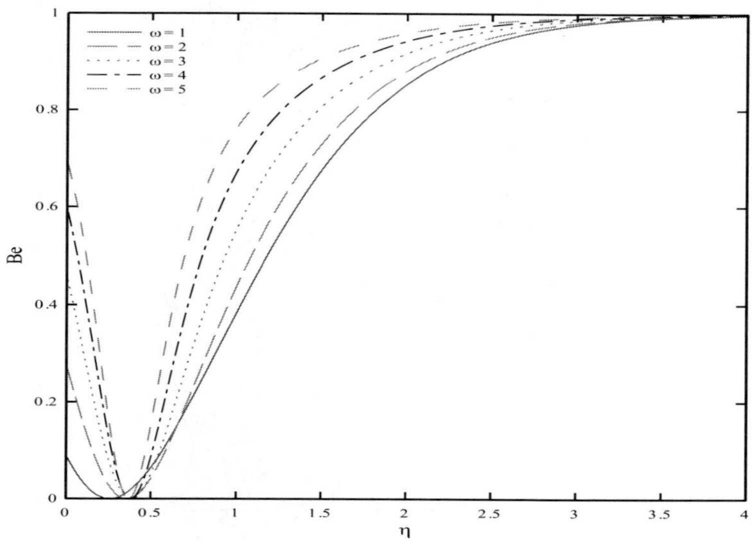

Figure 25. The variation of Bejan number, *Be* with rotation strength parameter for $Ec = Pr = 1,\ M = 2,\ Br = 3$ and $Re = 6.$

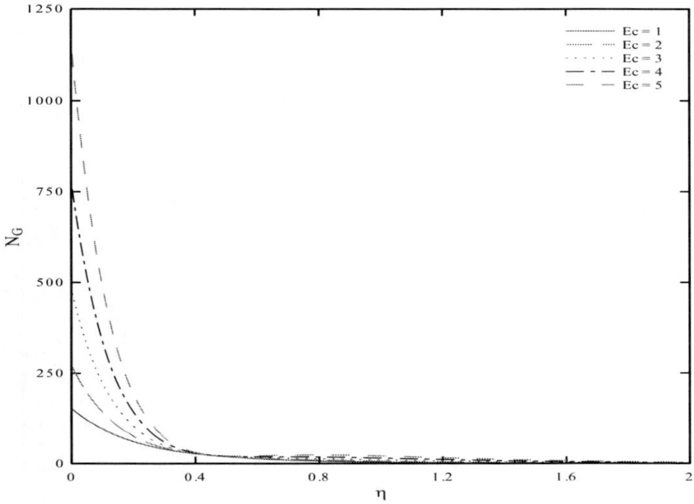

Figure 26. The variation of entropy generation number, N_G with Eckert (viscous heating) number for $Pr = 1$, $\omega = M = 2$, $Br = 4$ and $Re = 6$.

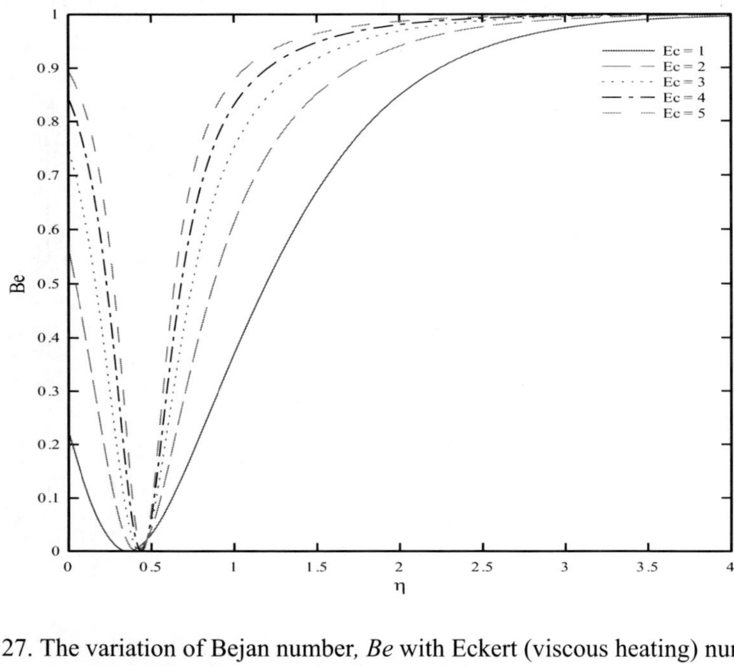

Figure 27. The variation of Bejan number, *Be* with Eckert (viscous heating) number for $Pr = 1$, $\omega = M = 2$, $Br = 4$ and $Re = 6$.

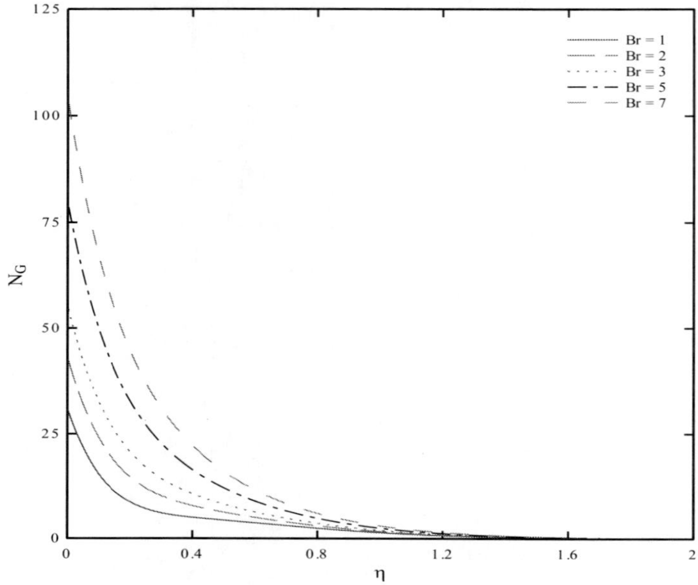

Figure 28. The variation of entropy generation number, N_G with Brinkman number for $\omega = 1$, $M = Ec = Pr = 2$ and $Re = 3$.

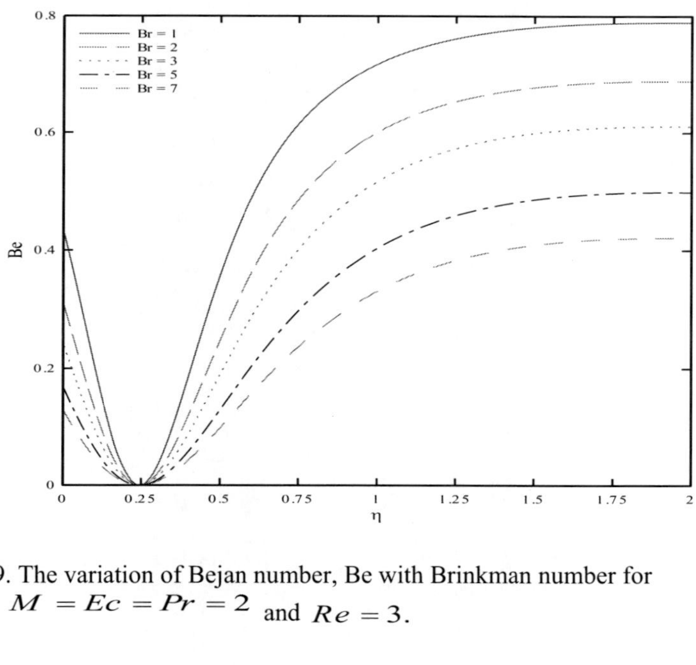

Figure 29. The variation of Bejan number, Be with Brinkman number for $\omega = 1$, $M = Ec = Pr = 2$ and $Re = 3$.

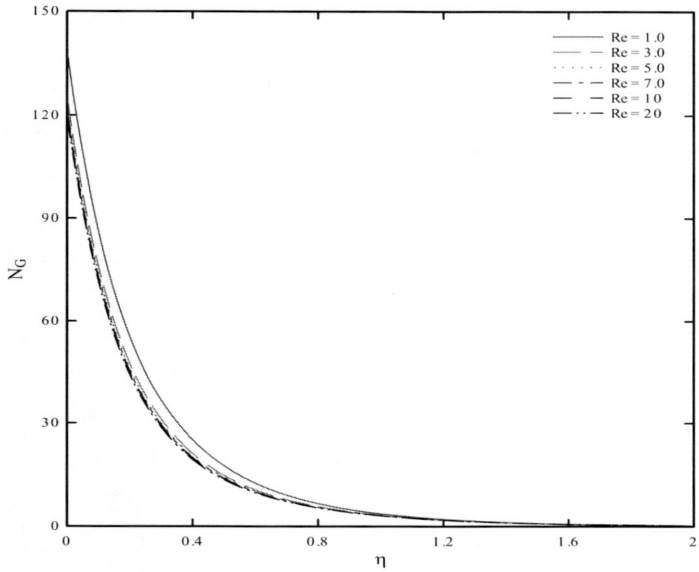

Figure 30. The variation of entropy generation number, N_G with Reynolds number for $Ec = Pr = 1$, $M = \omega = 2$ and $Br = 3$.

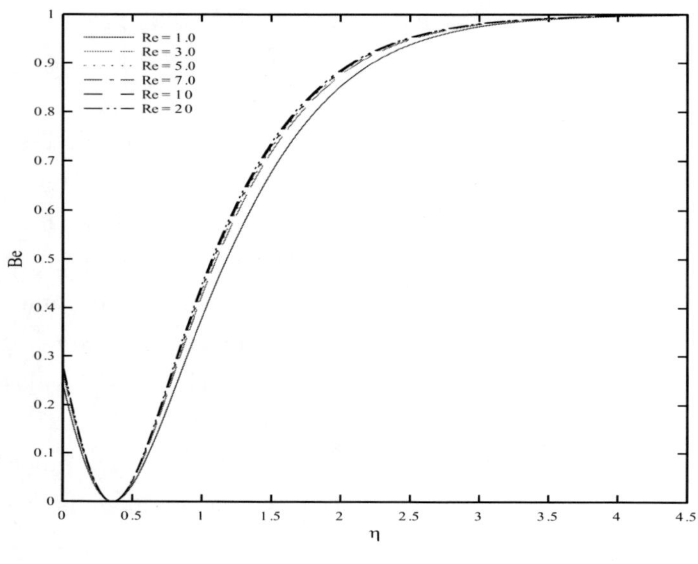

Figure 31. The variation of *Bejan number, Be* with Reynolds number for $Ec = Pr = 1$, $M = \omega = 2$ and $Br = 3$.

The method has been validated extensively [76-78]. To provide a systematic examination of the influence of the key thermophysical parameters on the momentum, heat and entropy in the swirling disk regime, we have conducted extensive computations which are depicted in Figures 18-31. All parameter data is provided in the relevant captions for each figure. Parameter values are feasible for EMP systems.

Magnetic parameter, $M = (\sigma B_0^2/\rho s)$ features in the radial and azimuthal (tangential) momentum conservation equations (32), (33) and also the energy conservation eqn. (35). It appears in the two momentum equations as $-MF$ and $-MG$, i.e., two linear retarding forces. These magnetic (Lorentzian) drag forces will therefore evidently *oppose* momentum in the radial and azimuthal directions, respectively. As such the momentum destroyed in these directions will be re-directed via the Von Karman swirl mechanism in the *axial* direction. In consistency with this, the radial velocity is strongly decelerated with an increase in magnetic interaction parameter, as is the azimuthal velocity. Conversely the axial velocity, is strongly enhanced with increasing magnetic parameter i.e., values become less negative. In all cases axial velocity is negative since flow will be drawn towards the disk surface in the negative z-direction as a result of the *swirl effect*. We further note that magnetic field effects are negated when $M = 0$ i.e., the flow is electrically non-conducting. The imposition of magnetic field necessitates extra work expenditure by the fluid which has to be dragged in the disk plane, against the Lorentzian retarding forces. This supplementary work is dissipated in the form of Joule (Ohmic) heating which manifests with a rise in temperature in the fluid, with increasing M values. The effect is prominent also since the Joule heating term features quadratic functions of the radial and azimuthal velocity, which although decreased individually with greater M values, still contribute to the term, $Ec\, M\,(F^2 + G^2)$. As such the overall effect on the temperature distribution is a significant increase and heating of the swirling boundary layer regime. Furthermore the classical temperature overshoot identified in numerous Von Karman hydromagnetic swirling flow studies is also computed. We further note that Prandtl number is set to 2 to simulate realistic water-based working fluids in nuclear by-pass MHD systems, Eckert number is high at 2 owing to high conversion rates of kinetic energy in the flow and the rotational strength parameter, $\omega = \Omega/s$, is prescribed as *unity* for which disk swirl is *equal* to the disk radial stretch. Although not illustrated, the influence of the rotation strength parameter (ω), is an accentuation in swirl due to the escalation in centrifugal forces. This draws fluid radially outwards in a *fan-mechanism* which is characteristic of Von Karman swirling flows. As a result the radial velocity is boosted considerably. Turbulence effects also contribute to enhancing both swirl and velocity overshoots, although the present study has only considered the laminar regime. Azimuthal (tangential) velocity is also found to be enhanced with an increase in rotational strength parameter. Since both radial and tangential momentum are escalated with increasing ω, there will be a depletion in axial momentum, owing to the fan effect. In accordance with this, axial velocity is markedly decreased with an increase in ω. The *maximum* axial velocity therefore corresponds to the case of *weakest* swirl i.e., $\omega. =1$. Temperature is significantly elevated with increasing rotational strength parameter. Clearly there is strong coupling with radial and azimuthal velocity to temperature function in the energy conservation equation (35), in the *viscous heating term*, $Ec\ M\ (F'^2 + G'^2)$ *and the Ohmic dissipation term*, $Ec\ M\ (F^2 + G^2)$. The boost in radial and tangential momentum with increasing swirl has a profound effect on escalating heat transfer from the disk surface into the boundary layer. Temperatures are therefore elevated in the entire boundary layer regime. Let us briefly consider viscous heating effects. As the value of Eckert number increases positively, this corresponds to an increase in convection currents from the disk to the fluid which transfers greater thermal energy to the boundary layer, manifesting in an increase in fluid temperatures. The opposite case of increasingly negative Eckert numbers implies *removal* of heat from the boundary layer regime to the disk via convection currents. As a result the disk temperature will be enhanced and fluid temperatures will be depressed. The intermediate case of $Ec = 0$ implies an absence of viscous dissipation where there is no conversion of kinetic energy to thermal energy and the classic profile of monotonic decay is observed. For positive Ec there will therefore be an increase in thermal boundary-layer thickness whereas with negative *Ec* the thermal boundary layer thickness will be reduced. It is also worth noting that increasing Prandtl number leads to a significant reduction in temperature. Prandtl number embodies the ratio of *viscous diffusion* to *thermal diffusion* in the boundary layer regime and furthermore expresses the ratio of the *product of specific heat capacity and dynamic viscosity*, to the *fluid thermal conductivity*. When *Pr* is high, viscous diffusion rate exceeds thermal diffusion rate. An increase in *Pr* from 1 through 3, 5, 7 to 9 greatly decreases temperatures. For $Pr < 1$, thermal diffusivity exceeds momentum diffusivity i.e., heat will diffuse faster than momentum. For $Pr = 1.0$, both the viscous and energy diffusion rates will be the same as will the thermal and velocity boundary layer thicknesses. For the case of $Pr = 1$ there is a clear temperature overshoot present near the disk surface. For higher values of *Pr* however this overshoot vanishes and profiles decay *monotonically* from the disk surface to the edge of the boundary layer. With lower *Pr* there is a more even distribution of temperature through the boundary layer. This diffusion of heat is depressed and effectively controlled with larger values of Prandtl number. Generally we consider the case of equal rotational strength and disk stretching i.e., $\omega = 1$. The cases of $\omega = 0$ i.e., *pure stretching* of the disk without rotation, and $\omega \to \infty$ i.e., *pure rotation* of the disk without stretching are extreme scenarios and not considered here. Figures 18-21 present colour visualizations of the *velocity contours* in all directions. As the radial coordinate increases, the primitive radial (u) and azimuthal (v) velocity components clearly increase. These velocity components are maximized near the surface of the disk i.e., at *low* values of *axial* coordinate (bottom right hand corner of both Figures 18, 19). The velocity vectors are shown to provide a good perspective of the fluid flow. In all cases although ω has been prescribed as 1, the disk stretch rate is very high at 10. Conversely in Figure 19 we observe that axial velocity component is maximized near the disk surface for all values of radial coordinate (lower red band). The axial velocity clearly decays as we depart from the surface and is minimized at greater distances from the disk ($z \sim 0.4$). Figure 20 distinctly shows the convergence of velocity vectors (u,w) towards the bottom right hand corner of the plot i.e., fluid is clearly drawn in a fan like mechanism outwards along the radial coordinate and in the negative axial direction, as elaborated earlier. This pattern typifies Von Karman swirling flow as elucidated by Zandbergen and Dijkstra [79]. Figures 21-32 illustrate the influence of the emerging thermophysical parameters on entropy generation number, (N_G) and Bejan number (*Be)* distributions, with radial coordinate. In each of these graphs, dimensionless temperature difference is set to unity $(\alpha = 1)$. The entropy generation number and Bejan number variation with changing *magnetic interaction parameter* (M) are presented in Figures 21, 22. As the magnetic interaction parameter increases, the amount of entropy generation increases considerably (it is minimized in the absence of a magnetic field, $M = 0$). This implies that in order to control the entropy which is generated from swirling flow on the disk surface, the value of magnetic interaction parameter should be *reduced*, an issue of interest in nuclear-MHD rotating disk propulsion. The dominant mechanism of irreversibility is due to fluid friction irreversibility, and is distant from of the disk surface for small values of magnetic interaction parameter. As the magnetic interaction parameter is elevated, the heat transfer mechanism dominates irreversibility. It should also be mentioned that a minimum point arises for Bejan number near the disk surface for any value of magnetic interaction parameter. At large values of M the Bejan number escalates (Figure 22) and after ascending very sharply from the disk surface converges asymptotically to the maximum allowable value of unity, a trend sustained into the free stream. In essence when entropy generation is *maximized*, the Bejan number is *minimized* and vice versa in the swirling flow regime. Figures 23-24 illustrate the effect of rotation strength parameter on the entropy generation and Bejan number distributions with radial coordinate. An increase in ω generates a similar but more pronounced response to that of increasing magnetic interaction parameter i.e., it causes a considerable increase in N_G near the disk surface, leading to a *maximum* entropy generation value always *at* the disk surface itself. The values computed for N_G are an order of magnitude greater with an increase in ω (Figure 23) as compared with Figure 21. Further from the disk N_G profiles decay sharply to vanish in the free stream. The Bejan number is generally minimized closer to the disk and maximized further from the disk, in direct contrast to the entropy generation number. However whereas in Figure 22 the Bejan number profiles at lower magnetic interaction parameters assume parabolic profiles with a maximum close to the disk surface, in Figure 24, the maxima are always attained for any value of rotation strength parameter, *far* from the disk surface. In all cases, the lowest value of Bejan number arises in the same vicinity as the maximum value of entropy generation number i.e., near the disk surface. For larger values of rotation strength parameter Bejan number achieves the asymptotic profile sooner. In the asymptotic state where the profiles all converge to unity in the free stream, the *heat transfer irreversibility* is *dominant.* Figures 25 and 26 depict the response of the entropy numbers to Eckert number. In Figure 25, we observe that increasing Eckert number significantly increases the entropy generation number, N_G increases, as the Eckert number increases. It should also be noted that at a certain distance from the disk surface $(\eta \cong 0.45)$, the amount of generated entropy converges to the same value for all Eckert numbers. The influence of Eckert number on the Bejan number (*Be*) profiles are similar to those induced by the rotational strength parameter (Figure 24). Bejan number decays from a high value at the disk surface to a minimum near the disk surface and then ascends rapidly to converge to unity in the free stream. A rise in *Ec* clearly enhances Bejan number. In all cases in Figures 25 and 26, we have considered positive *Ec* values i.e., heat is transferred from the disk surface to the fluid by convection currents. Figures 27-28 display the influence of Prandtl number on the entropy generation and Bejan number profiles. In Figure 27 strong monotonic decay in N_G values is seen from the disk surface to the free stream. N_G values are largest for the highest value of Prandtl number at the disk and for some distance into the boundary layer. Further from the disk a transition in response is observed and N_G values are minimized with largest *Pr* value. Further from the disk surface *heat transfer irreversibility* dominates. For large values of Prandtl number the *fluid friction irreversibility* dominates compared with heat transfer. Figure 28 shows that low Bejan numbers are produced close to the disk for $Pr > 1$. For low

Prandtl number (Pr = 1), however the Bejan number assumes a strong monotonic ascent from the disk surface and converges asymptotically to the maximum permissible value of unity in the free stream. Bejan number is clearly a maximum when entropy generation number is a minimum (Figure 3.27). We further note that the correct trends for entropy generation number and Bejan number have been achieved with DTM-Padé simulation in Figures 21-28, and concur with other published observations. Figures 29-30 illustrate the effect of Brinkman number on entropy generation and Bejan number distributions. As the Brinkman number increases, the entropy generation number *increases*. Brinkman number, $\mathrm{Br} = \left(\mu s^2 r^2 / k\,\Delta \mathrm{T}\right)$ and represents the ratio of direct heat conduction from the disk surface to the viscous heat generated by shear in the boundary layer. It is distinct from Eckert number which symbolizes the kinetic energy of flow to the boundary layer enthalpy difference. It is in fact more appropriate for industrial flows, whereas Eckert number often arises in high-speed aerodynamic flows [80]. Both parameters are quantifications of viscous dissipation effects. N_G profiles for all values of Br, descend from maxima at the disk surface to zero in the free stream. DTM-Padé solutions achieve excellent convergence. To decrease the entropy which is generated on the disk surface, the Brinkman number has to be reduced. Bejan number (Figure 30) is also generally *decreased* with an increase in Brinkman number. Profiles descend from an intermediate value at the disk surface to a minimum near the disk surface (η ~ 0.25) and thereafter *ascends monotonically* into the free stream. As the Brinkman number increases, the effect of *fluid friction irreversibility* is amplified on the disk surface. Further from the disk surface, the effect of *heat transfer irreversibility* becomes larger as the Brinkman number *decreases*. For high Br the effects of heat transfer and fluid friction irreversibilities are almost equal. Bejan number attains the *maximum* value (~ 0.8) only for the *minimum* value of Brinkman number (Br = 1) far from the disk surface. Effectively, Figures 29, 30 testify that the contribution of Brinkman number, as originally proposed and elucidated by Bejan [81] and is therefore of some significance in second law thermodynamic analysis. Figures 30-31 illustrate the effect of Reynolds number on the entropy generation number and Bejan number distributions. Increasing Reynolds number causes a decrease in N_G values which exhibit a gradual descent from the disk surface to the freestream. The N_G values *at* the disk surface are more closely clustered and less dispersed. Maximum entropy generation is achieved at the disk surface. Increasing Bejan numbers are also generated with an increase in Reynolds number. Although there is an additional plummet in *Be* values close to the disk surface, all profiles converge eventually to the theoretical limit of unity for any Reynolds number as we move away from the disk surface towards the edge of the boundary layer. Overall Bejan number is minimized at the same locations where entropy generation number peaks, and vice versa. At large distances from the disk the dominant effect is heat transfer irreversibility. The present computations have provided some further insight into the thermodynamics and fluid mechanics of proposed rotating disk MHD systems coupled with nuclear space propulsion engines, although these require further evaluation, in particular for *transient* and *magnetic induction* effects, aspects which are under consideration by the authors. DTM-Padé simulation holds excellent potential in analyzing such effects and can be relatively easily explored by propulsion engineers.

4. Biomimetic Magnetohydrodynamic Propulsion

The propulsion mechanisms studied thus far are for external thrust in spacecraft. However internal propulsion is a different issue, for example, the disposal of waste in space-stations, re-circulation of materials for combustion, renewable heat etc. For this type of application, biomimetic propulsion [82] is growing in interest. The study of flow dynamics of a fluid in a tube induced by a wave traveling on its wall, has many applications in various branches of science and engineering. The physical mechanism of the flow induced by the traveling wave is now well understood and is known as *peristaltic transport.* This mechanism is a natural cause of motion of fluids in the bodies of living creatures, and it frequently occurs in the organs such as oesophagus, ureters, intestines and arterioles. Interest of researchers in the peristaltic flows has increased substantially during the last few decades, initially in physiology but more recently in astronautical systems. A variety of models have been presented for the peristaltic flow of Newtonian fluid with constant viscosity and variable viscosity. Additionally non-Newtonian fluids such as power law fluid, generalized Newtonian fluids, third grade fluid, Jeffrey fluid, micropolar fluid and couple stress fluid through the gap between two concentric tubes, inner tube is rigid and outer tube is flexible, have also received attention. Representative studies include [83-92]. *Couple stress fluids* are a special type of a non-Newtonian fluid, whose particle sizes are taken into account. Since the classical continuum theory does not study the particle size effects, but a micro-continuum theory given by Stokes [93] has been considered to take into account the particle size effects. The Stokes micro-continuum theory is the generalization of the classical theory of fluids allowing for polar effects such as the presence of couple stresses, body couples, and an anti-symmetric stress tensor. In this fluid model, the couple stress effects are considered as a consequence of the action of a deforming body on its neighborhood. Some authors [94-96] have reported the peristaltic flow of the couple stress fluid through uniform and non-uniform channel/tube. They discussed the effects of couple stress parameter and magnetic field on pressure and friction force. Peristaltic propulsive magnetohydrodynamic flows have also received extensive attention in recent years. Many fluids possess an electrically-conducting nature in the presence of a transverse magnetic field. The so-called *magnetohydrodynamic body-force effect* can therefore be strategically exploited in a diverse range of applications including waste management, propulsion systems using conducting gases etc.

In the nuclear engineering industry, the need to transport plasma discharges and hot conducting waste slurry products (non-Newtonian in nature) efficiently and safely has also resulted in the exploitation of magneto-peristaltic pump devices. A particularly strong motivation for non-Newtonian hydromagnetic peristaltic flows with heat transfer is the mitigation of performance failures in such systems. For example unless systems are designed accurately, unanticipated defects and blockages in main tube can precipitate disconnections due to re-circulating "bolus" flows and induce a failure resulting in hazardous material being sprayed external to the pump [97]. A further application of peristaltic MHD thermal transport in rheological fluids is the control of sodium suspensions in fast breeder reactors for deep spacecraft mission designs [98].

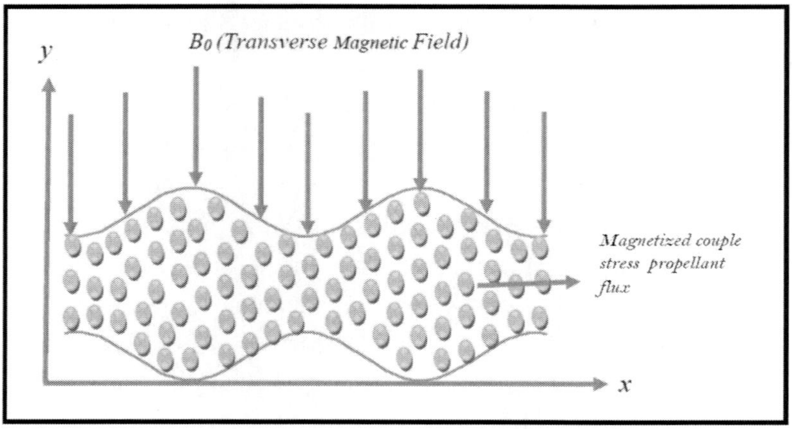

Figure 32. Bio-mimetic magnetohydrodynamic rheological propulsion model.

In aerodynamic boundary layer and internal chamber flows, MHD peristaltic waves also provide a very efficient control mechanism for delaying separation and mitigating detachment which can lead to inefficiencies and energy losses [99, 100]. Finally some very recent applications of hydromagnetic peristaltic flow include inductively coupled plasma mass spectrometry [101] and combustion spray flow control [102] where ionic states of burning products are regulated by the imposition of magnetic fields and flow discharge is further controlled with a peristaltic nozzle. In this section of our review we examine the *magnetized peristaltic flow of couple stress fluid through the gap between two coaxial channels containing an annular porous medium regime*. The couple stress model is employed to simulate micro-structural characteristics of gel-like working fluids. Magnetized and non-magnetic couple stress flows have also been considered by Zueco and Bég [103] for pulsating blood flow and Ghosh et al. [104] for magnetohydrodynamic and rotational flows in centrifugal blood purification systems. The porous medium achieves better thermal conduction in the propulsion and also allows flow control of peristaltic propulsion waves. The physical model is illustrated in Figure 32. The geometry of wall surface is simulated via the following equations:

$$\tilde{h}_1 = a_1, \tag{74}$$

$$\tilde{h}_2 = a_2 + b\sin\frac{2\pi}{\lambda}(\tilde{X} - c\tilde{t}), \tag{75}$$

where a_1, a_2 are the half width of the inner and outer channels, b is amplitude, λ is wavelength, c is the wave velocity and $\tilde{t}$ is time. Introducing a wave frame $(\tilde{\xi}, \tilde{\eta})$ moving velocity c away from the fixed frame ($\tilde{X}, \tilde{Y}$) by the transformation:

$$\tilde{\xi} = \tilde{X} - c\tilde{t}, \quad \tilde{\eta} = \tilde{Y}, \quad \tilde{u} = \tilde{U} - c, \quad \tilde{v} = \tilde{V}, \tag{76}$$

where $(\tilde{u}, \tilde{v})$ and $(\tilde{U}, \tilde{V})$ are the velocity components in the wave and fixed frame respectively. After using these transformations, and neglecting the body forces and the body couples, the continuity equations and the equations of motion for a couple stress fluid under the effect of magnetic fields, incorporating porous medium drag effects are:

$$\frac{\partial \tilde{u}}{\partial \tilde{\xi}} + \frac{\partial \tilde{v}}{\partial \tilde{\eta}} = 0, \tag{77}$$

$$\rho\left(\tilde{u}\frac{\partial \tilde{u}}{\partial \tilde{\xi}} + \tilde{v}\frac{\partial \tilde{u}}{\partial \tilde{\eta}}\right) = -\frac{\partial \tilde{p}}{\partial \tilde{\xi}} + \mu\nabla^2\tilde{u} - \mu_1\nabla^4\tilde{u} - \sigma B_0^2(\tilde{u} + c) - \frac{\mu}{\tilde{K}}(\tilde{u} + c), \tag{78}$$

$$\rho\left(\tilde{u}\frac{\partial \tilde{v}}{\partial \tilde{\xi}}+\tilde{v}\frac{\partial \tilde{v}}{\partial \tilde{\eta}}\right)=-\frac{\partial \tilde{p}}{\partial \tilde{\eta}}+\mu\nabla^2\tilde{v}-\mu_1\nabla^4\tilde{v}-\sigma B_0^2\tilde{v}-\frac{\mu}{\tilde{K}}\tilde{v}, \quad (79)$$

where $\rho,\tilde{u},\tilde{v},\tilde{\eta},\tilde{p},\mu,\mu_1,\sigma,B_0,\tilde{K}$ are the fluid density, axial velocity, transverse velocity, transverse coordinate, pressure, viscosity, constant associated with couple stress, electrical conductivity, transverse magnetic field, permeability parameter (hydraulic conductivity of porous medium), and $\nabla^2 \equiv \frac{\partial^2}{\partial \tilde{\xi}^2}+\frac{\partial^2}{\partial \tilde{\eta}^2}$, $\nabla^4=\nabla^2\nabla^2$. Proceeding with the analysis, we introduce the following dimensionless parameters:

$$\left.\begin{array}{l}\xi=\frac{\tilde{\xi}}{\lambda},\ \eta=\frac{\tilde{\eta}}{a_2},\ u=\frac{\tilde{u}}{c},\ v=\frac{\tilde{v}}{c\delta},\ h_1=\frac{\tilde{h}_1}{a_2}=\frac{a_1}{a_2}=\varepsilon,\ \phi=\frac{b}{a_2},\\ h_2=\frac{\tilde{h}_2}{a_2}=1+\phi\sin(2\pi\xi),\ p=\frac{\tilde{p}{a_2}^2}{\mu c\lambda},\ K=\frac{\tilde{K}}{a_2^2},\ \mathrm{Re}=\frac{\rho c a_2}{\mu},\\ \delta=\frac{a_2}{\lambda},\ \alpha=a_2\sqrt{\frac{\mu}{\mu_1}},\ M=\sqrt{\frac{\sigma}{\mu}}a_2B_0,\end{array}\right\} \quad (80)$$

where $\delta, \varepsilon, \phi, \mathrm{Re}, \alpha, M$ designate the wave number, ratio of half width of channels, amplitude ratio, Reynolds number, couple stress parameter and Hartmann (magnetohydrodynamic body force) number respectively. Utilizing the long wavelength and low Reynolds number approximation [105] in Eqns. (76)-(78) we obtain:

$$\frac{\partial u}{\partial \xi}+\frac{\partial v}{\partial \eta}=0, \quad (81)$$

$$\frac{\partial p}{\partial \xi}=\frac{\partial^2 u}{\partial \eta^2}-\frac{1}{\alpha^2}\frac{\partial^4 u}{\partial \eta^4}-M^2(u+1)-\frac{1}{K}(u+1), \quad (82)$$

$$\frac{\partial p}{\partial \eta} = 0. \tag{83}$$

The boundary conditions are prescribed thus:

No slip conditions, $u = 0 \ at \ \eta = h_1, \ u = 0 \ at \ \eta = h_2,$ (84)

Vanishing of couple stresses,

$$\frac{\partial^2 u}{\partial \eta^2} = 0 \ at \ \eta = h_1, \ \frac{\partial^2 u}{\partial \eta^2} = 0 \ at \ \eta = h_2. \tag{85}$$

We note that the Darcy model, as elucidated in detail by Bég et al. [106-108] is valid for low Reynolds number, *viscous dominated* flows, and therefore provides a consistent approach with the "low Reynolds number approximation," which neglects inertial effects. Effectively this adds a linear retarding body force term to the axial momentum eqn. (81). The solution of Eq. (81), with boundary conditions (83) and (84), is obtained readily as follows:

$$u = C_1 e^{m_1 \eta} + C_2 e^{-m_1 \eta} + C_3 e^{m_2 \eta} + C_4 e^{-m_2 \eta} - \left(1 + \frac{(\partial p/\partial \xi)}{N}\right), \tag{86}$$

where

$$m_1 = \sqrt{\frac{\alpha^2 + \alpha\sqrt{\alpha^2 - 4N}}{2}}, \tag{87}$$

$$m_2 = \sqrt{\frac{\alpha^2 - \alpha\sqrt{\alpha^2 - 4N}}{2}}, \tag{88}$$

$$C_1 = \frac{m_2^2(1+A)}{m_2^2 - m_1^2}\left\{\frac{e^{-m_1 h_1} - e^{-m_1 h_2}}{2\sinh(m_1(h_1 - h_2))}\right\}, \tag{89}$$

$$C_2 = \frac{m_2^2(1+A)}{m_2^2 - m_1^2}\left\{\frac{e^{m_1 h_1} - e^{m_1 h_2}}{2\sinh(m_1(h_2 - h_1))}\right\}, \tag{90}$$

$$C_3 = \frac{m_1^2(1+A)}{m_1^2 - m_2^2}\left\{\frac{e^{-m_2 h_1} - e^{-m_2 h_2}}{2\sinh(m_2(h_1 - h_2))}\right\}, \tag{91}$$

$$C_4 = \frac{m_1^2(1+A)}{m_1^2 - m_2^2}\left\{\frac{e^{m_2 h_1} - e^{m_2 h_2}}{2\sinh(m_2(h_2 - h_1))}\right\}, \tag{92}$$

$$A = -\left(1 + \frac{(\partial p/\partial \xi)}{N}\right) \text{ and } N = M^2 + \frac{1}{K}. \tag{93}$$

The corresponding stream function (as defined by the Cauchy-Riemann equations, $u = \dfrac{\partial \psi}{\partial \eta}$ and $v = -\dfrac{\partial \psi}{\partial \xi}$) is defined thus:

$$\psi = \frac{C_1}{m_1}(e^{m_1\eta} - e^{m_1 h_1}) - \frac{C_2}{m_1}(e^{-m_1\eta} - e^{-m_1 h_1}) + \frac{C_3}{m_2}(e^{m_2\eta} - e^{m_2 h_1}) - \frac{C_4}{m_2}(e^{-m_2\eta} - e^{-m_2 h_1}) + A(\eta - h_1) \tag{94}$$

The volume flow rate in the wave frame is given by:

$$q = \int_{h_1}^{h_2} u\,d\eta\,. \tag{95}$$

Using the appropriate terms in Eqn. (94), we get:

$$q = \frac{2(1+A)}{m_2^2 - m_1^2}\left[\frac{m_2^2}{m_1}\left\{\frac{\cosh(m_1(h_1-h_2))-1}{\sinh(m_1(h_1-h_2))}\right\} - \frac{m_1^2}{m_2}\left\{\frac{\cosh(m_2(h_1-h_2))-1}{\sinh(m_2(h_1-h_2))}\right\}\right] + A(h_2-h_1) \tag{96}$$

The volume flow rate in fixed frame is given by:

$$Q = \int_{h_1}^{h_2} U dY = \int_{h_1}^{h_2} (u+1) d\eta, \tag{97}$$

which, on integration yields:

$$Q = q + (h_2 - h_1). \tag{98}$$

Averaging volume flow rate along one time period, we have

$$\overline{Q} = \int_0^1 Q dt. \tag{99}$$

Further algebra yields:

$$\overline{Q} = q + 1 - \varepsilon. \tag{100}$$

It can also be shown that the *pressure gradient* takes the form:

$$\frac{\partial p}{\partial \xi} = -N\left\{1 + \frac{\overline{Q} - 1 + \varepsilon - f(\xi)}{(h_2 - h_1) + f(\xi)}\right\}, \tag{101}$$

where

$$f(\xi)=\frac{2}{(m_2^2-m_1^2)}\left[\frac{m_2^2}{m_1}\left\{\frac{\cosh(m_1(h_1-h_2))-1}{\sinh(m_1(h_1-h_2))}\right\}-\frac{m_1^2}{m_2}\left\{\frac{\cosh(m_2(h_1-h_2))-1}{\sinh(m_2(h_1-h_2))}\right\}\right] \tag{102}$$

The pressure rise (Δp), inner frictional force (F_{inn}) and outer frictional force (F_{out}), across one wavelength, are respectively computed using the expressions:

$$\Delta p=\int_0^1 \frac{\partial p}{\partial \xi}d\xi, \tag{103}$$

$$F_{inn}=\int_0^1 h_1\left(-\frac{\partial p}{\partial \xi}\right)d\xi, \tag{104}$$

$$F_{out}=\int_0^1 h_2\left(-\frac{\partial p}{\partial \xi}\right)d\xi. \tag{105}$$

The *maximum averaged flow rate* which defines the averaged flow rate at *zero* pressure (i.e. $\Delta p=0$), may be derived based on Eqn.(100) and Eq.(102) and assumes the form:

$$\overline{Q}_0=1-\varepsilon-\frac{1-\int_0^1 \frac{f(\xi)}{(h_2-h_1)+f(\xi)}d\xi}{\int_0^1 \frac{1}{(h_2-h_1)+f(\xi)}d\xi}. \tag{106}$$

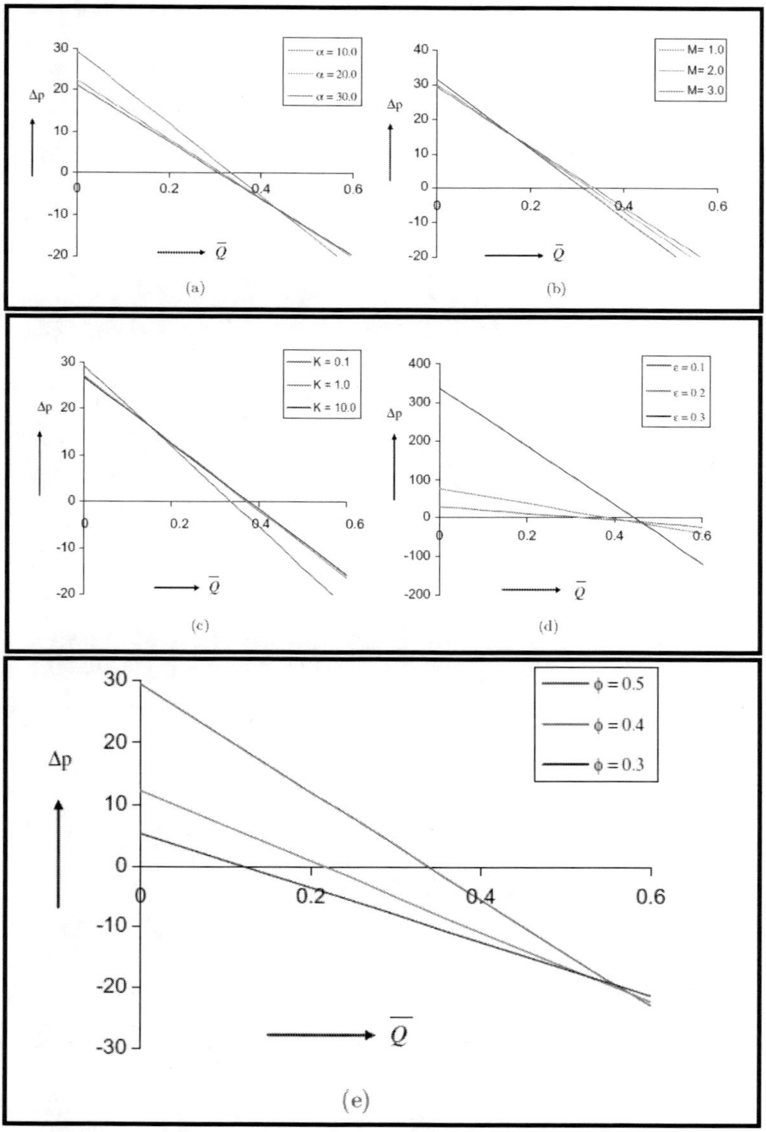

Figure 33. Pressure difference across one wavelength versus averaged flow rate for different parameter values: a) Couple stress parameter variation (α): ϕ *=0.5, M= 1.0, K = 0.1,* $\varepsilon = 0.1$ b)Hartmann number variation (*M*) ϕ *=0.5,* $\alpha = 10.0$*, K = 0.1,* $\varepsilon = 0.1$ c) permeability parameter (*K*) variation with ϕ *=0.5,* $\alpha = 10.0$*, M = 1.0,* $\varepsilon = 0.1$ d) Half width channel ratio (ε) variation with ϕ *=0.5,* $\alpha = 10.0$*, M = 1.0, K = 0.1* e) Wave amplitude variation (ϕ) with *M = 1.0, K = 0.1,* ε *=0.1,* $\alpha = 10.0$.

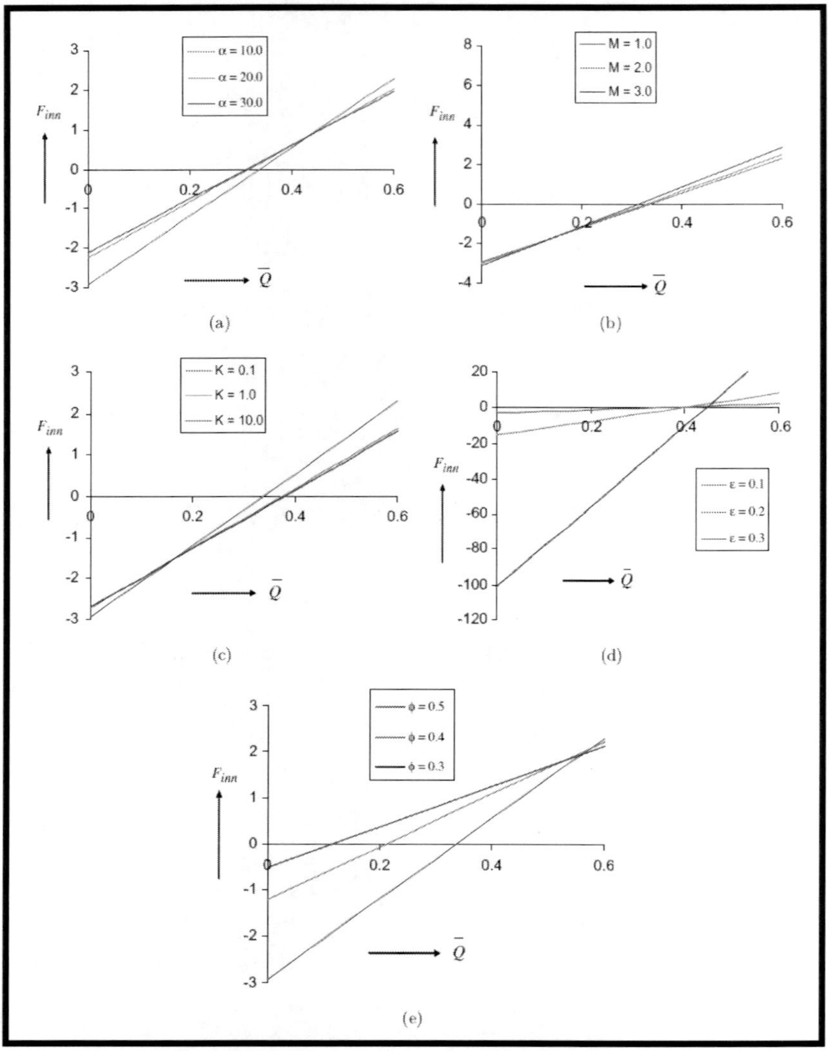

Figure 34. Friction force across one wavelength on wall of inner channel versus averaged flow rate for different parameter values: a) Couple stress parameter variation (α): $\phi = 0.5$, $M = 1.0$, $K = 0.1$, $\varepsilon = 0.1$ b) Hartmann number variation (M) $\phi = 0.5$, $\alpha = 10.0$, $K = 0.1$, $\varepsilon = 0.1$ c) permeability parameter variation (K) with $\phi = 0.5$, $\alpha = 10.0$, $M = 1.0$, $\varepsilon = 0.1$ d) Half width channel ratio (ε) variation with $\phi = 0.5$, $\alpha = 10.0$, $M = 1.0$, $K = 0.1$ e) Wave amplitude (ϕ) variation with $M = 1.0$, $K = 0.1$, $\varepsilon = 0.1$, $\alpha = 10.0$.

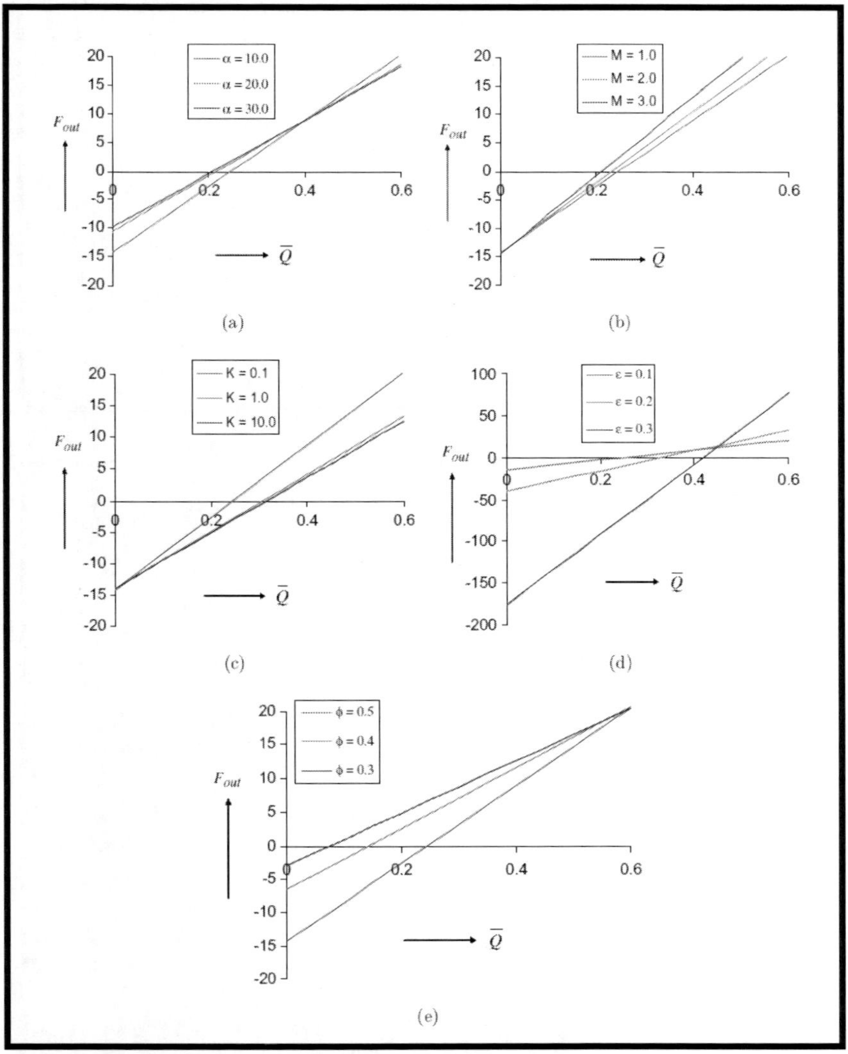

Figure 35. Friction force across one wavelength on wall of outer channel versus averaged flow rate for different parameter values: a) Couple stress parameter variation (α): $\phi = 0.5$, $M = 1.0$, $K = 0.1$, $\varepsilon = 0.1$ b)Hartmann number variation (M) $\phi = 0.5$, $\alpha = 10.0$, $K = 0.1$, $\varepsilon = 0.1$ c) permeability parameter variation (K) with $\phi = 0.5$, $\alpha = 10.0$, $M = 1.0$, $\varepsilon = 0.1$ d) Half width channel ratio (ε) variation with $\phi = 0.5$, $\alpha = 10.0$, $M = 1.0$, $K = 0.1$ e) Wave amplitude (ϕ) variation with $M = 1.0$, $K = 0.1$, $\varepsilon = 0.1$, $\alpha = 10.0$.

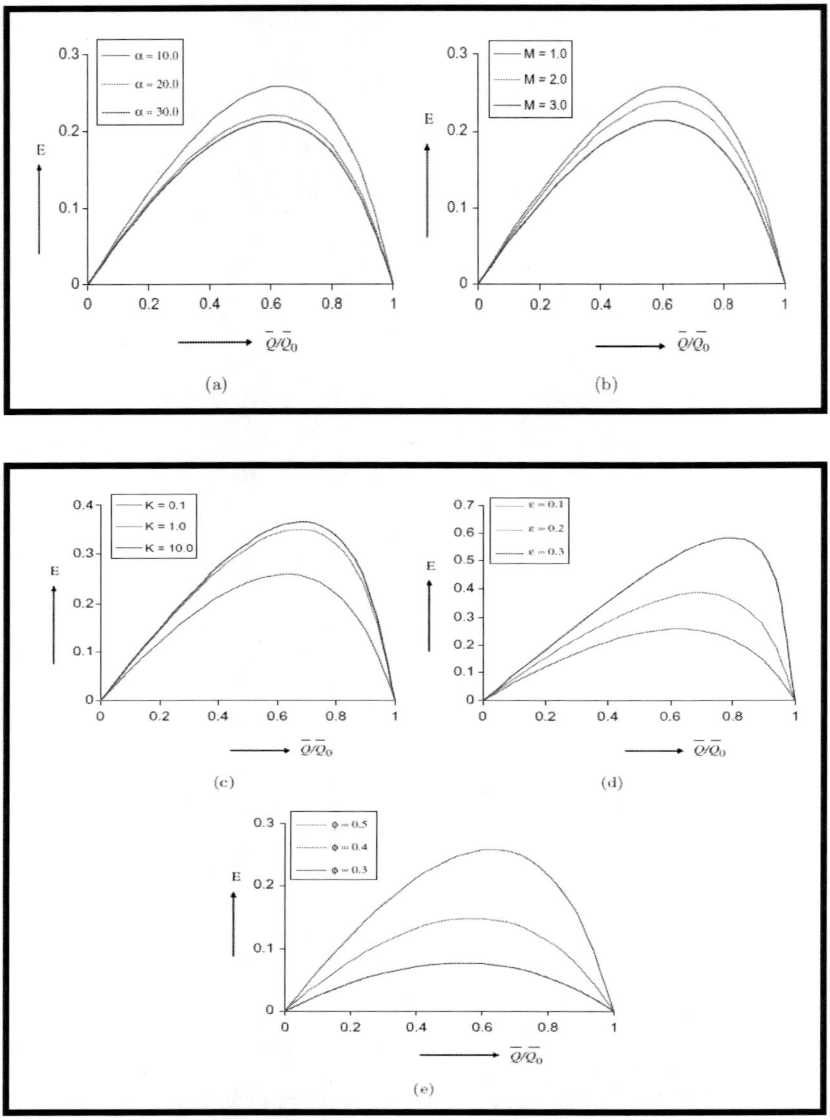

Figure 36. Mechanical efficiency versus averaged flow rate and maximum averaged flow rate for different parameter values: a) Couple stress parameter variation (α): $\phi = 0.5,\ M = 1.0,\ K = 0.1,\ \varepsilon = 0.1$ b) Hartmann number variation (M) $\phi = 0.5,\ \alpha = 10.0,\ K = 0.1,\ \varepsilon = 0.1$ c) permeability parameter variation (K) with $\phi = 0.5,\ \alpha = 10.0,\ M = 1.0,\ \varepsilon = 0.1$ d) Half width channel ratio (ε) variation with $\phi = 0.5,\ \alpha = 10.0,\ M = 1.0,\ K = 0.1$ e) Wave amplitude (ϕ) variation with $M = 1.0,\ K = 0.1,\ \varepsilon = 0.1,\ \alpha = 10.0$.

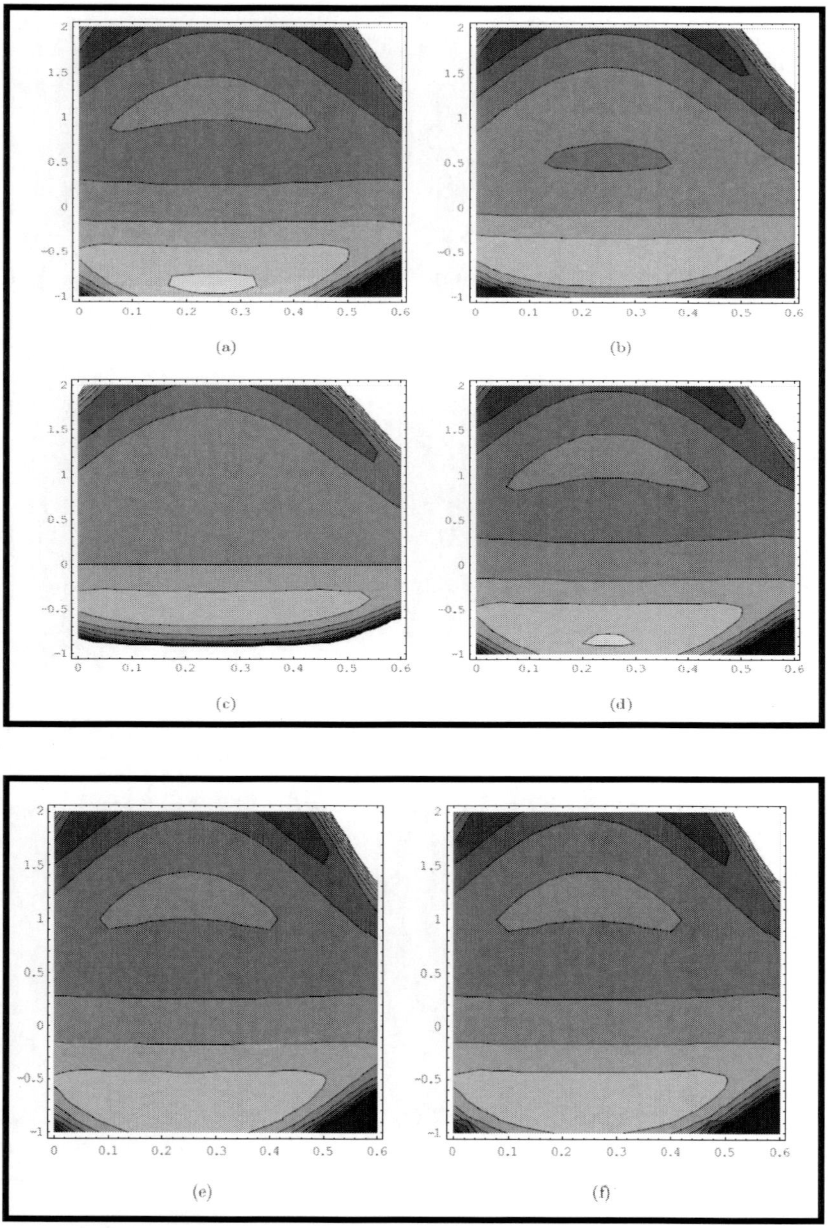

Figure 37. (Continued). Streamlines in the wave frame at ϕ =0.7,Q=0.6 for various parameter values: a) $\alpha = 1$, M= 1.0, K = 0.1, $\varepsilon = 0.1$ b) $\alpha = 2$, M=1, K = 0.1, $\varepsilon = 0.1$ c) $\alpha = 3$, M = 1.0,K=0.1, $\varepsilon = 0.1$ d) $\alpha = 1$, M = 2, K = 0.1, $\varepsilon = 0.1$ e) $\alpha = 1$, M = 3, K = 0.1, $\varepsilon = 0.1$ f) $\alpha = 1$, M = 4, K = 0.1, $\varepsilon = 0.1$.

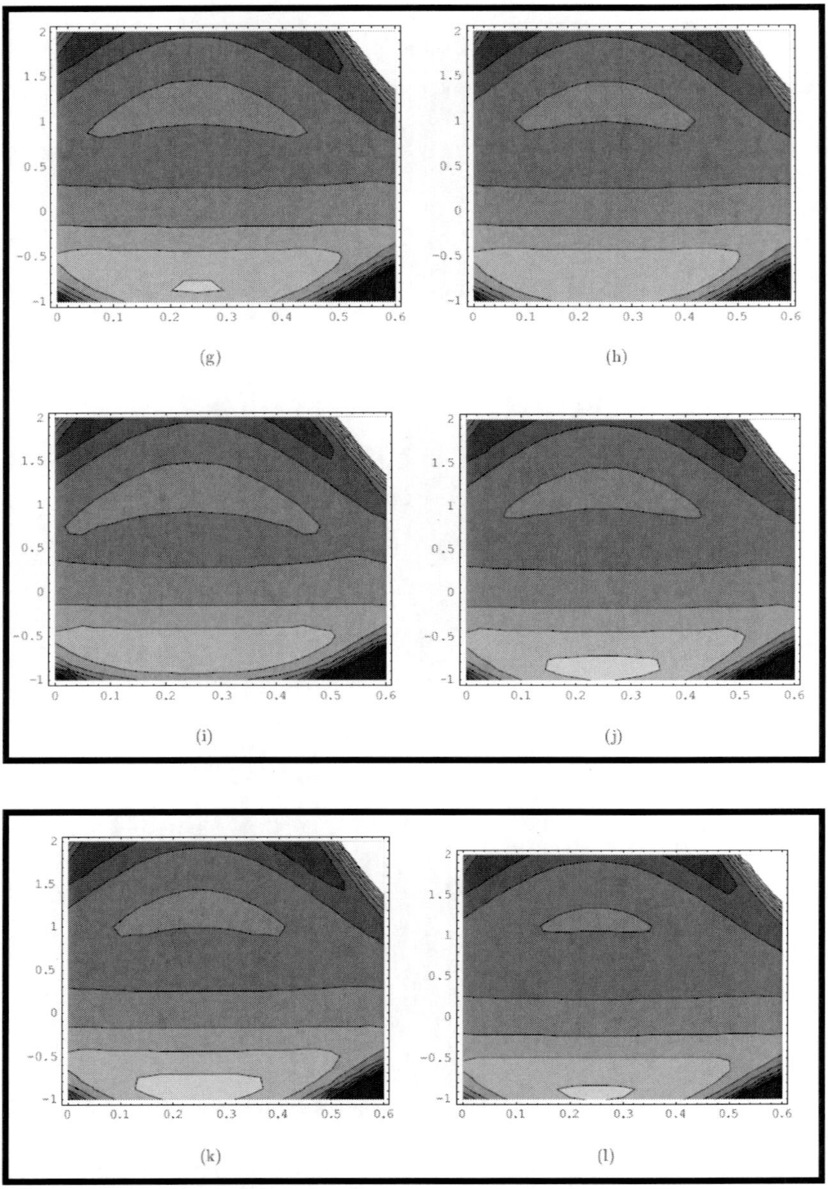

Figure 37. Streamlines in the wave frame at ϕ =0.7,Q=0.6 for various parameter values:g) α = 1, M= 1.0, K = 0.075, ε = 0.1 h) α = 1, M=1, K = 0.05, ε = 0.1 i) α = 1, M = 1.0,K=0.025, ε = 0.1 j) α = 1, M = 1, K = 0.1, ε = 0.09 k) α = 1, M = 1, K = 0.1, ε = 0.08 l) α = 1, M = 1, K = 0.1, ε = 0.06.

Mechanical efficiency is the ratio of the average rate per wavelength at which work is done by the moving fluid against a pressure head and the average rate at which the walls do work on the fluid, as expounded by Shapiro et al. [109]. It is defined as follows:

$$E = -\frac{\overline{Q}\Delta p}{\phi I}, \tag{107}$$

where

$$I = \int_0^1 \frac{\partial p}{\partial \xi} \sin(2\pi\xi) d\xi . \tag{108}$$

Using the closed-form expressions given earlier, numerical results are evaluated using Simpson's third rule and a code based on the **C**-programming language. Specifically we examine the influence of couple stress fluid characteristics on pressure rise per wavelength (Δp), inner frictional force (F_{inn}), outer frictional force (F_{out}) and mechanical efficiency (E). This is achieved via studying Figures 33(a-e)-36(a-e). Figures 33(a-e) present the variation of pressure across one wavelength against averaged flow rate under the effects of couple stress parameter (α), Hartmann number (M), permeability parameter (*K*), ratio of half width of channels (ε), and amplitude ratio (ϕ) respectively. The relation between them is found to be *linear*. Figures 34 (a-e) show the relation between inner frictional force and averaged flow rate under the effects of the same parameters. A relation between mechanical efficiency (E) and ratio of average flow rate and maximum average flow rate ($\overline{Q}/\overline{Q}_0$) is depicted in Figures 35 (a-e) under the influence of the relevant parameters. It is found that this relation is *nonlinear*. In Figure 33(a), the effect of couple stress parameter on pressure is illustrated. Inspection of this figure indicates that the pressure decreases with increase in magnitude of couple stress parameter. Couple stress fluids are much simpler than micropolar fluids [110], since they possess no microstructure at the *kinematic* level and therefore the kinematics of such fluids is *totally* described using the *velocity* field. The principal effect of couple stresses is therefore to introduce a *length-dependent effect* which is absent in the classical non-polar fluid dynamics (Navier-Stokes model). The parameter ($1/\alpha^2$) is equivalent to the ratio of the couple stress material constant (Π) to the product of length squared and dynamic viscosity i.e., $\Pi/L^2\mu$ [111]. This aspect has been emphasized by Cowin [112]. Increasing α serves to enhance the viscosity of the couple stress fluid and this manifests with a greater pressure build up. This effect has also been identified by Mekheimer [113] for *non-porous* MHD peristaltic flow and for pulsatile MHD couple stress flows by Chaturani and Upadhya [114] and Rathod and Tanveer [115]. The influence of Hartman number on pressure-averaged flow rate curve are shown in Figure 33(b). It is observed that pressure increases initially when Hartmann number increases, for low $\widetilde{Q}$. Increasing magnetic field strength generates a rise in Hartmann number, for constant electrical conductivity (σ), dynamic viscosity (μ) and outer channel half width (a_2). This acts to decelerate the flow owing to the retarding nature of the Lorentzian linear impedance term in the reduced momentum eqn. i.e., – $M^2\,(u+1)$ which leads to an *initial* escalation in pressure in the channel. A similar effect has been observed also for *micropolar* fluids in periodic flows under transverse magnetic fields as expounded by Bég et al. [116]. Magnetic field exerts a potent effect on the pressure distribution. Higher M values also maximize the negative pressure change sustained at higher averaged flow rate values ($\widetilde{Q}$). At lower M, for example $M = 1$, where the magnetic drag force and the viscous force in the regime are of the same order of magnitude, the negative pressure differences are however minimized. Therefore, overall to achieve improved performance at lower flow rates, in practical magneto-peristaltic pumps, higher magnetic field intensity (B_0) is needed i.e., better performance is achieved with stronger magnetic fields. Figure 33(c) illustrates the influence of permeability parameter, K, on pressure distribution. The values studied correspond to very sparsely packed media i.e., large void distributions in the regime, and reduced tortuosity [117]. Increasing K generates a concomitant reduction in the *Darcian impedance force* i.e., with lesser and lesser fibers present in the porous medium, the resistance to percolating propellant diminishes. This results in an initial decrease in pressure difference (Δp) at low $\tilde{Q}$ values; as averaged flow rate increases however, the pressure difference is escalated with increasing K values. With the lowest permeability studied (K = 0.1), which leads to a maximization in the Darcian drag force, -$1/K(u+1)$, the initial pressure difference at low $\tilde{Q}$ is conversely maximized whereas with a progressive increase in $\tilde{Q}$ it is strongly suppressed. Higher permeability yields a larger pressure difference at intermediate to high $\tilde{Q}$ values. The effect of a porous medium is therefore evidently quite strong and also different to that of a magnetic field. This effect has also been highlighted by Rao and Mishra [118] and Hayat et al. [119], although they have considered different non-Newtonian models, respectively, power-law and Maxwell viscoelastic models. To achieve higher pressure differences at lower operational flow rates, a combination of *strong* magnetic field and *low* permeability (denser packing of porous material) is required in clinical MHD devices. Figure 33(d) shows the effect of ratio of half-width of the channels (ε) on pressure difference. $\varepsilon = a_1/a_2$ i.e., ratio of inner channel half width to outer channel half width. It is evident that the pressure increases when width of the inner channel increases with fixed value of the width of the outer channel (higher ε). Conversely pressure difference is markedly depressed with lower ε values i.e., when the width of the outer channel increases with fixed value of the width of the inner channel. Comparing with Figures 33a-c, the half width ratio is seen to exert a much more dramatic effect on pressure distribution i.e., geometric modification in the MHD peristaltic device is the significant factor in controlling pressure, as compared with magnetic field, couple stress and permeability parameters, which are all hydrodynamic effects. The influence of these hydrodynamic parameters is however noticeable and in conjunction with the half width ratio, can be used to optimize performance. We further note that pressure differences generated with increasing ε are an order of magnitude higher than with variation in α, M or K. Figure 33(e) relates the effect of the peristaltic wave amplitude, ϕ,on pressure difference evolution with flow rate. As ϕ increases from 0.3 through 0.4 to 0.5, a distinct elevation arises in Δp, a trend which is sustained for practically all values of $\tilde{Q}$. The deviation between the profiles however is reduced as $\tilde{Q}$ increases, with profiles converging at very high $\tilde{Q}$ (~ 0.6). This pattern is absent from the other Figures 33a-d, where the profiles are found to intersect at a significantly lower value of $\tilde{Q}$. However the magnitudes of pressure difference even at high wave amplitude are found to be an order of magnitude lower than those generated with high half-width ratio, again testifying to the pre-eminence of *geometry* in controlling magnitudes of pressure. In order to illustrate the influence of all geometric and hydrodynamic parameters (α, M, K, ε and ϕ) on the inner frictional force (F_{inn}), Figures 34(a-e) are presented. Figures 35(a-e) correspond to the distribution of outer frictional force (F_{out}) again under the effect of these five "control" parameters. The results of frictional forces (inner and outer) are similar to that for the pressure in qualitative sense i.e., linear distributions. However they are both opposite in direction and quantitative sense. Inner friction force is found to be reduced markedly in Figure 34a, for increasing couple stress parameter (α), since as elucidated earlier, the increase in viscosity of the couple stress fluid with higher α will aid in the biofluid slipping more easily through the tube i.e., less resistance will be generated via the *shearing action* on the inner wall. Conversely F_{inn} will be reduced with in increasing Hartmann number (Figure 34b) at low values of $\tilde{Q}$ but slightly increased at higher values of $\tilde{Q}$. The negative values of F_{inn} correspond to flow reversal effects. With increasing permeability (K), as seen in Figure 34c, the magnitude of F_{inn} is continuously reduced for all values of averaged flow rate ($\tilde{Q}$). Increasing permeability therefore assists in the propulsion of the biofluid along the inner channel since a decreased Darcian drag force acts;

with lower permeability the inner force is enhanced owing to greater resistance from the porous medium fibers. To further refine the modeling of the porous medium influence on the inner channel force a *vorticity-diffusion* model may be required based on the Brinkman boundary friction approach. This has been studied in the context of micropolar flows by Bég et al. [120] and is currently being considered for couple stress fluids also. A massive reversal in F_{inn} is induced with increasing the geometric half width ratio parameter (ε) for most values of $\widetilde{Q}$, as shown in Figure 34d. Similarly amplitude parameter (ϕ) (Figure 34) significantly lowers F_{inn} values which are in this case, *several orders of magnitude* lower than those for Figure 3d. Figures 35a–e reveal that while similar trends are observed for the outer channel friction force (F_{out}), there is a considerable difference in magnitudes; for the effects of four of the control parameters (α, M, K and ϕ). Only in the case of the geometric half width ratio parameter, ε, are similar values computed for both inner and outer channel frictional force (Figures 34d and 35d, respectively). The variations of mechanical efficiency (E) against the ratio of averaged flow rate and maximum averaged flow rate ($\overline{Q}/\overline{Q}_0$) for different parameters (α, M, K, ε and ϕ) are displayed in Figures 36(a-e). It is observed that that E first increases to attain a maximum value and then decreases to zero. It is found that the efficiency decreases with increase the magnitude of couple stress parameter and Hartmann number whereas it increases with increase in the magnitude of permeability parameter, ratio of width channel and amplitude. Porous medium, geometric and peristaltic wave effects are therefore the most beneficial for enhancing mechanical efficiency of *peristaltic pumping*, whereas couple stress and magnetic field effects are found to be counter-productive. The stream lines in wave frame are drawn under the influence of pertinent parameters through the Figures 37(a-l) to study a significant phenomenon known as trapping. It refers to closed circulating streamlines that exist at very high flow rates and when occlusions are very large. The trapped region moves with the *wave-velocity* i.e., *celerity of the peristaltic wave*. The effects of couple stress parameter, Hartmann number, permeability parameter and ratio of width channels on trapping are shown in Figures 37(a-c), Figures 37(a, d-f), Figures 37(a, g-i) and Figures 37(a, j-l) respectively. It is revealed that the size of the trapped region reduces with increasing the magnitude of couple stress parameter and Hartmann number. The size of the trapped region increases with increasing the magnitude of permeability parameter and the ratio of the width of channels. Trapping is therefore accentuated with a more permeable regime and inhibited largely with increasing magnetic field strength and couple stress fluid viscosity. An important outcome of this model is that hydro-mechanical efficiency of the peristaltic pumping is observed to decrease with increasing magnitude of α and M whereas it increases with increase in the magnitude of K, ε and ϕ. The present model considers only the Lorentzian magnetohydrodynamic drag effect in peristaltic non-Newtonian flow. Important modification to the model would entail *Hall current* effects and *oblique* magnetic fields [121-124].

CONCLUSION

Three different nonlinear mathematical models have been presented for different electro-magnetic propulsion (EMP) systems, namely electrohydrodynamic ion drag pumping, swirling magneto-plasma nuclear disk engines and biomimetic peristaltic magnetohydrodynamic propulsion. In each case the appropriate multi-physics models have been non-dimensionalized and simplified for the geometry considered. The resulting nonlinear boundary value problems have been solved with Chebyschev spectral methodology, differential transform methodology with Padé approximants and Mathematica numerical integration methods. Many different phenomena have been studied in these propulsion systems including electrical conduction, Lorentzian body force, heat transfer, non-Newtonian rheology of working fluids, entropy generation, porous media and deformable chamber walls. Interesting results have been addressed which expound on the different challenges of each system and the efficiency attainable. Much work remains in all the systems considered before the implementation of any or all of the designs described. However they do promise to address many problems which arise with conventional propulsion which is not feasible for large-scale planetary missions of the 21st century and beyond. Furthermore combinations of some of the mechanisms described are also being explored e.g., electrohydrodynamic peristaltic propulsion, and the results of such investigations will be communicated imminently [125].

REFERENCES

[1] R. G. Jahn, *Physics of Electric Propulsion*, MacGraw-Hill, New York (1968).

[2] R. G. Jahn and E. Y. Choueiri, *Electric Propulsion, Encyclopedia of Physical Science, and Technology*, Third Edition, Volume 5, Academic Press, New York (2002).

[3] M. J. Patterson and S. W. Benson, NEXT Ion Propulsion System Development Status and Performance, AIAA–2007–5199, *43rd AIAA/ASME/SAE/ASEE Joint Propulsion Conference and Exhibit*, Cincinnati, OH, July 8–11 (2007).

[4] R. J. Melcher, *Continuum Electromechanics*, MIT Press, USA (1981).

[5] G. I. Taylor, Disintegration of water droplets in an electric field, *Proc. Royal Society A* 280, 383 (1964).

[6] C. T. Wilson and G. I. Taylor, The bursting of soap bubbles in a uniform electric field, *Proc. Cambridge Philos. Soc.,* 22 (5): 728 (1925).

[7] A. A. Martins and M. J. Pinheiro, Fluidic electrodynamics: Approach to electromagnetic propulsion, *Physics of Fluids,* 21, 9, 097103-097103-7 (2009).

[8] O. Anwar Bég, J. Zueco, Tasveer A. Bég and R. Bhargava, Network simulation of the electrohydrodynamic ion drag energy pump with electrical Reynolds number, slip and source effects, *Int. J. Applied Mathematics and Mechanics,* 6, 78-95 (2010).

[9] Trefethen L. N., *Spectral Methods in MATLAB*. SIAM: Philadelphia, (2000).

[10] Canuto C., Hussaini M., Quarteroni A. and Zang T. *Spectral Methods in Fluid Dynamics*, Springer*:* Berlin (1993).

[11] Balachandar, S. and Streett, C. L., Numerical simulation of transition in a rotating disk flow, Instability and transition*; Proc. Workshop,* Hampton, Virginia, USA, May 15-June 9, Vol. 2 (A91-31301 12-34) (1989).

[12] Raad, P. E. and A. Karageorghis, A Chebyshev Spectral Collocation Method for the solution of the Reynolds equation of lubrication, *J. Computational Physics,* 106, 42-51 (1993).

[13] M. Avgousti, B. Liu and A. N. Beris, Spectral methods for the viscoelastic time-dependent flow equations with applications to Taylor–Couette flow, *Int. J. Numerical Methods Fluids,* 17, 49-74 (1993).

[14] W. S. Don, Numerical study of pseudospectral methods in shock wave applications, *J. Computational Physics,* 110, 103-111 (1994).

[15] W. Dauksher and A. F. Emery, Accuracy in modeling the acoustic wave equation with Chebyshev spectral finite elements, *Finite Elements in Analysis and Design,* 26, 115-128 (1997).

[16] S. Selvarajan, E. G. Tulapurkarab and V. V. Ramc, A numerical study of flow through wavy-walled channels, *Int. J. Numer. Meth. Fluids,* 26, 519-531 (1998).

[17] M. Arias-Zugasti and D. E. Rosner, Multicomponent fuel droplet vaporization and combustion using spectral theory for a continuous mixture, *Combustion and Flame,* 135, 271-284 (2003).

[18] Y. S. Sun and B. W. Li, Chebyshev collocation spectral approach for combined radiation and conduction heat transfer in one-dimensional semi-transparent medium with graded index, *Int. J. Heat Mass Transfer,* 53, 1491-1497 (2010).

[19] L. Huang, C. O. Ng and A. T. Chwang, A Fourier–Chebyshev collocation method for the mass transport in a layer of power-law fluid mud, *Computer Methods in Applied Mechanics and Engineering,* 195, 1136-1153 (2006).

[20] Md Mainul Hoque; Md Mahmud Alam; M. Ferdows, O. Anwar Bég, Numerical simulation of Dean number and curvature effects on magneto-biofluid flow through a curved conduit, *Proc. IMECHE-Part H; J. Engineering in Medicine,* 227 (11):1155-70 (2013).

[21] O. Anwar Bég, M. M. Hoque, M. Wahiuzzaman, M. Mahmud and M. Ferdows, Spectral numerical simulation of laminar magneto-physiological Dean flow, *J. Mechanics in Medicine and Biology,* 14 (3) (2014) (18 pages).

[22] M. Wahiduzzaman, M. M. Alam, M. Ferdows, S. Sivasankaran and O. Anwar Bég, Spectral numerical study of non-isothermal flow in a rotating rectangular cross-section duct, *Int. J. Applied Math. Mechanics,* 10(2): 61 – 78 (2014).

[23] O. Anwar Bég, S. S. Motsa, M. N. Islam and M. Lockwood, Pseudo-spectral and variational iteration simulation of exothermically-reacting Rivlin-Ericksen viscoelastic flow and heat transfer in a rocket propulsion duct, *Computational Thermal Sciences,* 6, 1, 1-12 (2014).

[24] O. Anwar Bég, J. Zueco, T. A. Bég and M. N. Islam, PSPICE electro-thermal computational simulation of the electrohydro-dynamic propulsion, Tech. Rep- AERO-4/2009, Sheffield Hallam University, *Aerospace Engineering,* May (2009).

[25] M. Y. Hussaini, and T. A. Zang, Spectral methods in fluid dynamics, *Annual Review of Fluid Mechanics,* 19, 339-367 (1987).

[26] V. Singhal and S. V. Garimella, Influence of bulk fluid velocity on the efficiency of electrohydrodynamic pumping, *ASME J. Fluids* Eng. 127, 484-494 (2005).

[27] Yanada, H. Y. Toshihiro, Asai, Y. and Terashita, Y., Measurement and numerical simulation of ion drag pump characteristics, *JSME J. Fluid Science Technology,* 5, 617-631 (2010).

[28] O. M. Stuetzer, Magnetohydrodynamics and electrohydrodynamics, Phys. *Fluids,* 5, 534-544 (1962).

[29] A. A. Martins, Modelling an improved positive corona thruster and actuator, *J. Electrostatics,* 71, 61-67 (2013).

[30] E. Choueiri, A critical history of electric propulsion: the first 50 years (1906 - 1956), *AIAA J. Propulsion and Power*, 20, 193-203 (2004).

[31] A. Bejan, *Entropy Generation through Heat and Fluid Flow*, Wiley, New York (1982).

[32] A. Bejan, *Entropy Generation Minimization: The Method of Thermodynamic Optimization of Finite-Size Systems and Finite-Time Processes*, CRC Press, Florida (1996).

[33] A. Bejan, A study of entropy generation in fundamental convective heat transfer, *ASME J. Heat Transfer.* 101, 718-725 (1979).

[34] S. Aïboud, S. Saouli, Entropy analysis for viscoelastic magneto-hydrodynamic flow over a stretching surface, *Int. J. Non-Linear Mechanics.* 45, 482-489 (2010).

[35] E. Abu-Nada, Entropy generation due to heat and fluid flow in backward facing step flow with various expansion ratios, *Int. J. Exergy,* 3, 419-435 (2006).

[36] L. Erbay, M. Ercan, B. Sülüs, M. Yalçÿn, Entropy generation during fluid flow between two parallel plates with moving bottom plate, *Entropy,* 5, 506-518 (2003).

[37] M. Al-Odat, R. Damseh, M.d. Al-Nimr, Effect of magnetic field on entropy generation due to laminar forced convection past a horizontal flat plate, *Entropy,* 6, 293-303 (2004).

[38] S. Aïboud, S. Saouli, Second law analysis of viscoelastic fluid over a stretching sheet subject to a transverse magnetic field with heat and mass transfer, *Entropy,* 12, 1867-1884 (2010).

[39] A. Z. Sahin, The effect of variable viscosity on the entropy generation and pumping power in a laminar fluid flow through a duct subjected to constant heat flux, *Heat and Mass Transfer,* 35, 499-506 (1999).

[40] G. Ibáñez, S. Cuevas, Entropy generation minimization of a MHD (magnetohydrodynamic) flow in a micro-channel, *Energy*, 35, 4149-4155 (2010).

[41] Makinde, O. D. and O. Anwar Bég, On inherent irreversibility in a reactive hydromagnetic channel flow, *J. Thermal Science*, 19, 1-8 (2010).

[42] A. Arikoglu, I. Ozkol, G. Komurgoz, Effect of slip on entropy generation in a single rotating disk in MHD flow, *Applied Energy,* 85, 1225-1236 (2008).

[43] H. Salas, S. Cuevas, M. L. Haro, Entropy generation analysis of magnetohydrodynamic induction devices, *J. Physics D: Applied Physics,* 32 (1999) 2605–2608.

[44] O. Anwar Bég, S. K. Ghosh and Tasveer A. Bég, *Applied Magnetofluid Dynamics: Theory and Computation, Lambert* Academic, Germany (21011).

[45] P. Intani, T. Sasaki, T. Kikuchi and N. Harada, Analysis of disk AC MHD generator performance by finite element method, *J. Plasma Fusion Res.,* 9 (2010), 580585.

[46] Ghosh, S. K., Bég, O. Anwar, J. Zueco and Prasad, V. R., Transient hydromagnetic flow in a rotating channel permeated by an inclined magnetic field with magnetic induction and Maxwell displacement current effects, ZAMP: *J. Applied Mathematics and Physics*, 61 (2010) 147-169.

[47] V. R. Prasad, B. Vasu, O. Anwar Bég and R. Parshad, Unsteady free convection heat and mass transfer in a Walters-B viscoelastic flow past a semi-infinite vertical plate: a numerical study, *Thermal Science-International Scientific J.,* 15 (2011) S291-S305.

[48] O. Anwar Bég, V. R. Prasad, B. Vasu, N. Bhaskar Reddy, Q. Li and R. Bhargava, Free convection heat and mass transfer from an isothermal sphere to a micropolar regime with Soret/Dufour effects, *Int. J. Heat and Mass Transfer,* 54 (2011) 9-18.

[49] Harbi A. Daud, Qinling Li, O. Anwar Bég and S. A. A. AbdulGhani, Numerical investigations of wall-bounded turbulence, *Proc. I Mech E. Part C: J. Mechanical Engineering Science,* 225 (2011) 1163-1174.

[50] D. Tripathi, O. Anwar Bég and J. Curiel-Sosa, Homotopy semi-numerical simulation of peristaltic flow of generalised Oldroyd-B

fluids with slip effects, *Computer Methods Biomechanics Biomedical Eng.*, 17, 433-442 (2014).

[51] M. M. Rashidi, M. Keimanesh, O. Anwar Bég and T. K. Hung, Magneto-hydrodynamic biorheological transport phenomena in a porous medium: A simulation of magnetic blood flow control and filtration, *Int. J. Numerical Methods in Biomedical Engineering*, 27 (2011) 805–821.

[52] T. Hayat, R. Naza, S. Asghar and S. Mesloub, Soret–Dufour effects on three-dimensional flow of third grade fluid, *Nuclear Engineering and Design* 243 (2012) 1- 14.

[53] S. Anghaie, Prospect of nuclear power MHD conversion, *Proc. 6th Int. Symp. Propulsion for Space Transportation of the 21st Century*, Palais des Congres, Versailes, France, May 14-17 (2002).

[54] S. Chandrasekhar, *Hydromagnetic and Hydrodynamic Stability*, Oxford, UK (1960).

[55] J. P. Thibault, F. Joussellin, A. Alemany, A. Dupas, The possible use of a metal-gas MHD energy converter in space, *Acta Astronautica,* 10 (1983) 587-589.

[56] H. Yamasaki, S. Kabashima, S. Shioda and Y. Okuno, Unsteady discharge and fluid flow in a closed-cycle disk MHD generator, *AIAA J. Propulsion and Power,* 4 (1988) 61-67.

[57] M. Turkyilmazoglu, MHD fluid flow and heat transfer due to a stretching rotating disk, *Int. J. Thermal Sciences*, 51 (2012) 195-201.

[58] T. Fang, Flow over a stretchable disk, *Phys. Fluids*, 19 (2007) 128105-128110.

[59] O. Anwar Bég, Lik Sim, J. Zueco and R. Bhargava, Numerical study of magnetohydrodynamic viscous plasma flow in rotating porous media with Hall currents and inclined magnetic field influence, *Communications in Nonlinear Science and Numerical Simulation*, 15 (2010) 345-359.

[60] T. Von Kármán, Über läminare und turbulence Reibung, *ZAMM,* 1 (1921) 233-252.

[61] M. M. Rashidi, N. Laraqi, S. M. Sadri, A novel analytical solution of mixed convection about an inclined flat plate embedded in a porous

medium using the DTM-Padé, *Int. J. Thermal Sciences*. 49 (2010) 2405-2412.

[62] J. K. Zhou, *Differential Transformation and Its Applications for Electrical Circuits*, Huazhong University Press, Wuhan, China. (1986).

[63] M. M. Rashidi, E. Erfani, A new analytical study of MHD stagnation-point flow in porous media with heat transfer, *Computers & Fluids*. 40 (2011) 172-178.

[64] R. Attarnejad and A. Shahba, Application of differential transform method in free vibration analysis of rotating non-primsatic beams, *World Appl. Sci. J.,* 4, 441-448 (2008).

[65] Cha'o-Kuang Chen, Hsin-Yi Lai, Chin-Chia Liu, Numerical analysis of entropy generation in mixed convection flow with viscous dissipation effects in vertical channel, *Int. Comm. Heat and Mass Transfer,* 38, 285–290 (2011).

[66] M. M. Rashidi, M. Keimanesh, Using Differential Transform Method and Padé approximant for solving MHD flow in a laminar liquid film from a horizontal stretching surface, *Mathematical Problems in Engineering*. 2010 (2010) 1-14.

[67] M. M. Rashidi, E. Erfani, O. Anwar Bég and S. K. Ghosh, Modified Differential Transform Method (DTM) simulation of hydromagnetic multi-physical flow phenomena from a rotating disk, *World J. Mechanics,* 1, 217-230 (2011).

[68] M. M. Rashidi, S. A. Mohimanian Pour, A novel analytical solution of heat transfer of a micropolar fluid through a porous medium with radiation by DTM-Padé, *Heat Transfer-Asian Research*. 39, 575-589 (2010).

[69] M. M. Rashidi, O. Anwar Bég, M. Asadi and M. T. Rastegari, DTM-Padé modeling of natural convective boundary layer flow of a nanofluid past a vertical surface, *Int. J. Thermal and Environmental Engineering,* 4 (2012) 13-24.

[70] M. M. Rashidi, N. Rahimzadeh, M. Ferdows, Md. Jashim Uddin and O. Anwar Bég, Group theory and differential transform analysis of mixed convective heat and mass transfer from a horizontal surface

with chemical reaction effects, *Chemical Engineering Communications,* 199 (2012),1012-1043.

[71] M. M. Rashidi, A. Basiri Parsa, O. Anwar Bég, L. Shamekhi, S. M. Sadri and Tasveer A. Bég, Parametric analysis of entropy generation in magneto-hemodynamic flow in a semi-porous channel with OHAM and DTM, *Applied Bionics and Biomechanics,* 11, 47-60 (2014).

[72] G. A. Baker, P. R. Graves-Morris, Padé, *Approximants: Basic Theory.* Addison-Wesley, (1981).

[73] O. Anwar Bég, R. Bhargava and M. M. Rashidi, *Numerical Simulation in Micropolar Fluid Dynamics*, Lambert, Germany (2011).

[74] C. Brezinski and M. Redivo Zaglia, *Extrapolation Methods. Theory and Practice.* North-Holland (1991).

[75] M. Keimanesh, M. M. Rashidi, O. Anwar Bég, and A. Maligno, Numerical simulation of heat transfer in non-Newtonian flow through a porous channel using multi-step differential transform method (MDTM), *Int. J. Applied Math. Mechanics,* 10(1): 42 – 61 (2014).

[76] M. M. Rashidi, B. Rostami, S. Abbasbandy and O. Anwar Bég, Homotopy semi-numerical simulation of Soret and Dufour effects on magneto-convection heat and mass transfer from a transpiring stretching surface, *Int. J. Applied Math. Mechanics,* 10(1): 88 - 120, (2014).

[77] M. M. Rashidi, O. Anwar Bég, B. Rostami and L. Osmond, DTM-Padé simulation of stagnation-point nanofluid mechanics, *Int. J. Applied Mathematics and Mechanics,* 9(3): 1 - 29, (2013).

[78] P. J. Zandbergen and D. Dijkstra, D., Von Kármán swirling flows. *Ann. Rev. Fluid Mech.* 19, 465-491 (1987).

[79] J. D. Anderson, *Hypersonic and High Temperature Gas Dynamics*, AIAA, New York (2000).

[80] A. Bejan, *Advanced Engineering Thermodynamics*, 3rd ed., Wiley, Hoboken (2006).

[81] J. O. Dabiri, Optimal vortex formation as a unifying principle in biological propulsion, *Ann. Rev. Fluid Mechanics* 41: 17-33 (2009).

[82] T. Hayat, N. Ali, S. Asghar and A. M. Siddiqui, Exact peristaltic flow in tubes with an endoscope. *Applied Mathematics and Computation*, 182 (2006) 359-368.

[83] T. Hayat, N. Ahmad and N. Ali, Effects of endoscope and magnetic field on the peristalsis involving Jeffrey fluid, *Communication in Nonlinear Science and Numerical Simulation,* 13 (2008) 1581-1591.

[84] T. Hayat and N. Ali, Effects of an endoscope on peristaltic flow of a micropolar fluid, *Mathematical and Computer Modelling*, 48 (2008) 721-733.

[85] Kh. S. Mekheimer and Y. Abd-elmaboud, Peristaltic flow of a couple stress fluid in an annulus: Application of an endoscope, *Physica A,* 387 (2008) 2403-2415.

[86] Kh. S. Mekheimer and Y. Abd-Elmaboud, The influence of heat transfer and magnetic field on peristaltic transport of a Newtonian fluid in vertical annulus: Application of an endoscope*, Physics Letters A,* 372 (2008)1657-1665.

[87] A. Afsar, N. Ali and T. Hayat, Effects of an endoscope on the peristaltic transport through a porous medium, *J. Porous Media*, 86 (2008) 477-786.

[88] T. Hayat, E. Momoniat and F. M. Mahomed, Peristaltic MHD flow of third grade fluid with an endoscope and variable viscosity*, J. Nonlinear Mathematical Physics,* 15 (2008), 91-104.

[89] A. E. H. A. El-Naby and A. E. M. El-Misery, Effects of an endoscope and generalized Newtonian fluid on peristaltic motion, *Applied Mathematics and Computation,* 128 (2002) 19-35.

[90] A. E. H. A. El Naby, A. E. M. El Misery and I. I. El Shamy, Effects of an endoscope and fluid with variable viscosity on peristaltic motion, *Applied Mathematics and Computation*, 158 (2004) 497-511.

[91] A. M. El Misery, A. E. H. A. El Naby and A. H. El Nagar, Effects of a fluid with variable viscosity and an endoscope on peristaltic motion, *J. Phys. Soc. Jpn.,* 72 (2003) 89-93.

[92] V. K. Stokes, Couple stress fluids, *Phys. Fluids*, 9 (1966) 1709-1715.

[93] L. M. Srivastava, Peristaltic transport of a couple stress fluid, *Rheologica Acta*, 25 (1986) 638-641.

[94] E. F. El Shehawey and Kh. S. Mekheimer, Couple stress in peristaltic transport of fluids, *J. Phys. D: Appl. Phys.,* 27 (1994) 1183-1170.

[95] E. F. El Shehawey and W. El-Sebaei, Couple stress in peristaltic transport of a magneto- fluid, *Physica Scripta*, 64 (2001) 401-409.

[96] R. Johanson and K. Simmons, Causal analysis for sodium hydroxide (NaOH) spray from peristaltic pump tube failure, *J. Failure Analysis and Prevention,* 5, (2005) 62-66.

[97] J. Ishimoto and S. Kamiyama, Wave propagation in two-phase flow of magnetic fluid under a non-uniform magnetic field, *Nuclear Engineering and Design,* 175, (1997) 87-96.

[98] R. C. M. Madhan, J. Rahel and J. R. Roth, Aerodynamic flow control using paraelectric and peristaltic acceleration of a one atmosphere uniform glow discharge plasma, *31st IEEE International Conference on Plasma Science,* 28 June-1 July, 245 (2004).

[99] J. Reece Roth, Aerodynamic flow acceleration using paraelectric and peristaltic electrohydrodynamic effects of a one atmosphere uniform glow discharge plasma, *Physics of Plasmas,* 10 (2003) 2117-2125.

[100] E. Björn, T. Jonsson and D. Goitom, The origin of peristaltic pump interference noise harmonics in inductively coupled plasma mass spectrometry, *Journal of Analytical and Atomistic Spectrometry,* 17 (2002) 1390-1393.

[101] X. L. Sun, A. I. Y. Tok, S. L. Lim, F. Y. C. Boey, C. W. Kang and H. W. Ng, Combustion-aided suspension plasma spraying of Y2O3 nanoparticles:, Synthesis and modeling, *Journal of Applied Physics,* 103 (2008) 1-13.

[102] J. Zueco and O. Anwar Bég, Network numerical simulation applied to pulsatile non-Newtonian flow through a channel with couple stress and wall mass flux effects, *Int. J. Applied Mathematics and Mechanics,* 5, (2009) 1-16.

[103] S. K. Ghosh, O. Anwar Bég, R. Bhargava, S. Rawat, and T. A. Bég, Mathematical modelling of transient magnetohydrodynamic couple stress fluid flow in a rotating channel, *Int. J. Applied Mathematics and Mechanics* 6 (2010) 23 - 45.

[104] Y. C. Fung, *Biomechanics*, McGraw Hill, New York (1981).

[105] O. Anwar Bég, Bakier, A. Y. and V. R. Prasad, Numerical study of free convection magnetohydrodynamic heat and mass transfer from a stretching surface to a saturated porous medium with Soret and Dufour effects, *Computational Materials Science,* 46, 57-65 (2009).

[106] O. Anwar Bég, J. Zueco, H. S. Takhar, T. A. Bég and A. Sajid, Transient non-linear optically-thick radiative-convective double-diffusive boundary layers in a Darcian porous medium adjacent to an impulsively started surface: network simulation solutions, *Communications in Nonlinear Science and Numerical Simulation.,* 14, 3856-3866 (2009).

[107] O. Anwar Bég, S. K. Ghosh and M. Narahari, Mathematical modelling of oscillatory MHD Couette flow in a rotating highly permeable medium permeated by an oblique magnetic field, *Chemical Engineering Communications,* 198, 235-254 (2010).

[108] A. H. Shapiro, M. Y. Jaffrin and S. L. Weinberg, Peristaltic pumping with long wavelengths at low Reynolds number, *J. Fluid Mech.*, 35 (1969) 669-675.

[109] A. C. Eringen, *Micro-continuum Field Theories, Volume 2- Fluent Media*, Springer, New York (2001).

[110] O. Anwar Bég, *Couple stress rheological models for gel propellants, Aerospace Engineering*, Tech. Report AERO-7, May, Engineering and Mathematics Department, Sheffield Hallam University, UK (2010).

[111] S. C. Cowin, The theory of polar fluids, *Adv. Appl. Mech.* 14 (1974) 279-347.

[112] Kh. S. Mekheimer, Effect of the induced magnetic field on the peristaltic flow of a couple stress fluid, *Physics Letters A*, 372 (2008) 4271-4278.

[113] P. Chaturani P and V. S. Upadhya, Pulsatile flow of a couple stress fluid through circular tubes with applications to blood flow, *Biorheology,* 15 (1978) 193-201.

[114] V. P. Rathod and S. Tanveer, Pulsatile flow of couple stress fluid through a porous medium with periodic body acceleration and magnetic field, *Bull. Malays. Math. Sci. Soc.* 32 (2009) 245–259.

[115] S. Rawat, R. Bhargava, Renu Bhargava and O. Anwar Bég, Transient magneto-micropolar free convection heat and mass transfer through a non-Darcy porous medium channel with variable thermal conductivity and heat source effects, *Proc. IMechE Part C-J. Mechanical Engineering Science,* 223 (2009) 2341-2355.

[116] O. Anwar Bég and O. D. Makinde, Viscoelastic flow and species transfer in a Darcian high-permeability channel, *Petroleum Science and Engineering,* 76, 93–99 (2011).

[117] A. R. Rao and M. Mishra Peristaltic transport of a power-law fluid in a porous tube, *J. Non-Newtonian Fluid Mech.,* 121 (2004) 163–174.

[118] T. Hayat, N. Ali and S. Asghar, Hall effects on peristaltic flow of a Maxwell fluid in a porous medium, *Physics Letters A,* 363 (2007) 397–403.

[119] O. Anwar Bég, Prasad, H. S. Takhar, V. M. Soundalgekar, Thermo-convective flow in an isotropic, homogenous medium using Brinkman's model: Numerical Study*, Int. J. Numerical Methods in Heat and Fluid Flow,* 8, 59-89 (1998).

[120] J. Zueco, P. Eguía, E. Granada, J. L. Míguez and O. Anwar Bég, An electrical network for the numerical solution of transient MHD Couette flow of a dusty fluid: Effects of variable properties and Hall current, *Int. Comm. Heat Mass Transfer,* 37 (2010) 1432–1439.

[121] I. Pop, V. M. Soundalgekar, The Hall effect on an unsteady flow due to a rotating infinite disc, *Nuclear Engineering and Design*, 44 (1977) 309-314.

[122] O. Anwar Bég, J. Zueco and H. S. Takhar, Unsteady magnetohydrodynamic Hartmann–Couette flow and heat transfer in a Darcian channel with Hall current, ionslip, viscous and Joule

heating effects: *Network numerical solutions, Communications in Nonlinear Science Numerical Simulation,* 14, 1082-1097 (2009).

[123] J. Zueco, O. Anwar Bég and L. M. Lopez-Ochoa, Non-linear transient hydromagnetic partially ionised dissipative Couette flow in a non-Darcian porous medium channel with Hall, ionslip and Joule heating effects, *Progress in Computational Fluid Dynamics*, 11, 116-129 (2011).

In: Mathematical Modeling ISBN: 978-1-53613-162-8
Editors: Seth Sparks and Bill Willis

Chapter 2

MATHEMATICAL MODELING OF TRACER KINETICS FOR MEDICAL APPLICATIONS

Kenya Murase*, PhD
Department of Medical Physics and Engineering,
Graduate School of Medicine, Osaka University, Osaka, Japan

ABSTRACT

A tracer kinetic model is a mathematical description of the transport processes of the tracer within the tissue. Various methods have been proposed for the quantification of kinetic parameters including perfusion. These methods can be roughly classified into model-dependent and model-independent methods.

The model-dependent method usually employs compartment models and generally requires specific knowledge about the vascular and compartmental architecture. The kinetic parameters are generally estimated using the nonlinear least-squares (NLSQ) method. Although the NLSQ method is a powerful tool for estimating kinetic parameters, it

* Address correspondence to: Kenya Murase, Dr. Med. Sci., Dr. Eng.; Department of Medical Physics and Engineering, Division of Medical Technology and Science, Faculty of Health Science, Graduate School of Medicine, Osaka University, 1-7 Yamadaoka, Suita, Osaka 565-0871, Japan. Tel & Fax: (81)-6-6879-2571, Email: murase@sahs.med.osaka-u.ac.jp.

is not efficient in terms of computation time. Furthermore, poor initial guesses result in long computation times or occasionally failure to converge on a good fit. As an alternative approach, the linear least-squares (LLSQ) method has been introduced, which is computationally efficient and does not need initial guesses of the parameters. The graphical method has also been proposed.

The model-independent method yields kinetic parameters for a tissue region described as a "black box" and thus does not provide insight into its internal structure. For quantification of kinetic parameters using the model-independent method, deconvolution between the measured tissue residue curve and the vascular input function is needed to obtain the tissue response function. Deconvolution, however, is an ill-posed problem especially when there exists statistical noise. Thus, it is necessary to convert an ill-posed problem to a well-posed one using the so-called "regularization". Of the regularization methods, the truncated singular value decomposition (TSVD) is a popular method for obtaining regularized solutions. In this chapter, the use of TSVD in the model-independent method is introduced. The application of spectral analysis is also introduced. This method provides a spectrum of the kinetic components which are involved in the regional uptake and partitioning of tracer from the blood to the tissue and allows the tissue response function to be derived with minimal modeling assumptions.

The kinetic models employed in compartment analysis can be classified into lumped parameter and distributed parameter models. In the lumped parameter model, the compartments are assumed to be well-mixed, that is, tracer movement is sufficiently fast and distributes evenly throughout the compartments, such that tracer concentration is only a function of time, but not space. On the other hand, a distributed parameter model attempts to account for tracer concentration gradient within the compartments, such that concentration can be a function of both time and space. One of the simple distributed parameter models is the tissue homogeneity model. In this chapter, adiabatic approximation to the tissue homogeneity model is also described.

Finally, an empirical mathematical model is introduced. This model has been developed to describe contrast uptake and washout behavior without use of vascular input functions. This approach does not require making assumptions about underlying physiology or anatomy. The primary disadvantage of this approach, however, is that the parameters obtained by this approach do not correspond directly to identifiable physiological or anatomic features.

This chapter will be helpful for gaining a deeper understanding of mathematical modeling of tracer kinetics for medical applications.

Keywords: tracer kinetic modeling, model-dependent method, model-independent method, compartment model, tissue homogeneity model, empirical mathematical model

1. Introduction

Tracer kinetic modeling has its roots in pharmacokinetics that arose from the need to predict the concentration of drugs in blood and tissues [1]. Much of pharmacology has been driven by the need to arrive at optimal drug dosing, frequency of dosing, and drug concentration safety windows [1]. Therefore, pharmacologists have developed methods to make predictions based on blood time-concentration curves obtained by sampling the blood concentrations of a drug. Fitting of pharmacokinetic data is usually performed under various model assumptions. These approaches have limited capability to determine the dose of the drug delivered to the target in tissue. Positron emission tomography (PET) or single photon emission computed tomography (SPECT) provides the unique opportunity to make available not only the blood time-activity data but also regional tissue time-activity data [2]. These data allow a dramatic improvement in performing pharmacokinetics. Pharmacokinetic models when used with tracer levels of drug or tracer are referred to as "tracer kinetic models".

Dynamic contrast-enhanced computed tomography (DCE-CT) acquires X-ray CT images serially after the administration of intravenous contrast agent. Recently, DCE-CT has become a major imaging technique and has been widely used for diagnosing normal tissue, benign and malignant tumors, as well as for monitoring treatment effectiveness during and after therapy [3]. In addition to its easy accessibility in most clinical facilities, DCE-CT can provide quantification of the haemodynamics of lesions, if combined with appropriate tracer kinetic analysis [4]. One of the advantages of using CT for tracer kinetic analysis is that the relationship between the concentration of the administered iodinated contrast agent in the tissue and the attenuation of X-rays is almost always linear [4]. On the other hand, the major disadvantages of using CT are its potentially high radiation exposure due to repetitive scanning and the need for iodinated contrast agent that can be nephrotoxic and allergenic [5]. Recently, however, several studies have been performed to reduce these drawbacks in DCE-CT [6].

Dynamic contrast-enhanced magnetic resonance imaging (DCE-MRI) is the acquisition of serial MR images before, during, and after an intravenous administration of contrast agent. This technique has become increasingly important for estimating tissue physiological parameters such as perfusion, capillary permeability, and the volume of extravascular-extracellular space [7, 8]. The revealed kinetics of wash-in and wash-out of contrast agent are used to characterize masses as malignant or benign, and to determine whether a tumor is responding to therapy [7, 8]. Compared with methods involving the use of PET or SPECT, the quantification of physiological parameters by DCE-MRI offers several advantages, such as improved spatial resolution, no exposure of the patient to ionizing radiation, and the ability to combine morphological and functional information during a single imaging session [8].

In this chapter, the basic theory and mathematical modeling of tracer kinetic analysis and the practical methods for estimating kinetic parameters using PET, DCE-CT or DCE-MRI will be described.

2. Tracer Kinetic Analysis

2.1. Model-Dependent Method and Model-Independent Method

A tracer kinetic model is a mathematical description of the transport processes of the tracer within the tissue [9]. Recently, various methods have been proposed for the quantification of kinetic parameters including perfusion [10, 11]. These methods can be roughly classified into two groups; one is a model-dependent method and the other is a model-independent method.

For a model-dependent method, compartment models are usually employed as tracer kinetic models, and kinetic parameters are quantified using compartment analysis based on the assumption that tissues can be kinetically represented by a set of interacting subcompartments that represent distinguishable states of an administered tracer [4].

The kinetic models employed in compartment analysis can be classified into lumped parameter and distributed parameter models. The compartments are used to describe the physical distribution spaces of the tracer in the tissue. Conventionally, these compartments are assumed to be well-mixed, that is, tracer movement is sufficiently fast and distributes evenly throughout the compartments, such that tracer concentration is only a function of time, but not space. Thus, conventional compartment models are called lumped parameter models. On the other hand, a distributed parameter model attempts to account for tracer concentration gradient within the compartments, such that concentration can be a function of both time and space [9].

The model-dependent method generally requires specific knowledge about the vascular and compartmental architecture, the volumes of the compartments comprising the model, and the tracer-dependent permeability of compartment barriers. The assumptions in these models are generally valid under normal conditions, but for pathological conditions it is often difficult to independently verify which model parameters changed significantly, in particular when the model involves a large number of parameters, or if the sensitivity ranges for the different parameters overlap significantly [12].

In contrast to the model-dependent method, a model-independent method yields kinetic parameters for a tissue region described as a "black box" and thus does not provide insight into its internal structure [4]. For the quantification of kinetic parameters such as perfusion using the model-independent method, deconvolution between the measured tissue residue curve and the arterial input function is needed to obtain the tissue response function, from which the perfusion can be quantified based on the central volume principle [13]. Thus, the model-independent method is also called "deconvolution analysis".

Generally, model-independent deconvolution analysis has been applied to the data obtained by intravascular tracers [14]. Recently, the accuracy of model-independent deconvolution analysis when applied to the data obtained by extravascular contrast agents, i.e., the effect of the leakage of contrast agent from the intravascular space to the extravascular space has been studied [15].

To investigate the effect of the leakage of contrast agent from the intravascular space to the extravascular space, it is necessary to use a model in which the concentration of contrast agent within the tissue is treated as a function of both time and space, i.e., distributed parameter models [11]. The distributed parameter models describe tracer concentration in the vascular compartment as a function of both time and position along the capillary and are generally believed to be more realistic than the lumped parameter models. One of the simple distributed parameter models is the tissue homogeneity model, which was initially developed by Johnson and Wilson [16]. This model is based on both convective and diffusive transport of contrast agent in capillaries represented together as a single cylinder, and is composed of the intravascular space surrounded by the extravascular space. The two spaces are separated by a permeable barrier. St. Lawrence and Lee [17] derived an approximate closed-form solution to the tissue homogeneity model using the adiabatic approximation. The details of the adiabatic approximation to the tissue homogeneity model will be described afterwards.

2.2. Model-Dependent Method

In the model-dependent method, compartment models are usually employed, and the kinetic parameters are quantified using compartment analysis based on the assumption that tissues can be kinetically represented by a set of interacting subcompartments [4].

2.2.1. Two-Compartment Model

The majority of clinical studies have used a simple two-compartment model illustrated in Figure 1. This model is known as "extended Tofts model" [18]. Unless an independent estimate of blood flow or capillary permeability is essential, this model may be an appropriate choice for analysis of DCE-CT or DCE-MRI data [19]. This model assumes that the contrast agent resides in and exchanges between two compartments in the tissue: an intravascular plasma space (IVPS) and an extravascular-extracellular space (EES) (Figure 1). When using this model, the differential equation describing the kinetic behavior of the contrast agent in the tissue of interest is given by

$$v_e \frac{dC_e(t)}{dt} = K^{trans} \cdot \left[C_p(t) - C_e(t)\right], \tag{1}$$

where $C_e(t)$ and $C_p(t)$ are the concentrations of the contrast agent at time t in the EES and IVPS, respectively, and v_e is the distribution volume of contrast agent in the EES per unit volume of tissue, i.e., $v_e = V_e/V_t$, where V_e and V_t denote the volumes of the EES and tissue, respectively. K^{trans} is the volume transfer constant of contrast agent between the IVPS and EES. When we assume that the contrast agent does not pass into the intracellular space of the tissue, the concentration of the contrast agent in the tissue ($C_t(t)$) is given by $C_t(t) = v_p C_p(t) + v_e C_e(t)$, where v_p is the volume of the IVPS per unit volume of tissue, i.e., $v_p = V_p/V_t$, where V_p denotes the volume of the IVPS. Using this equation and by rearrangement of Eq. (1), we then obtain the following differential equation:

$$\frac{d[C_t(t) - v_p C_p(t)]}{dt} = K^{trans} C_p(t) - k_{ep} \cdot \left[C_t(t) - v_p C_p(t)\right], \tag{2}$$

where $k_{ep} = K^{trans}/v_e$, which represents the transfer rate constant between the IVPS and EES. In accord with recent standardization [20], K^{trans} is given by $K^{trans} = EF_p/V_t = EF_b(1 - Hct)/V_t$, where F_p and

F_b are plasma and whole blood flow, respectively, and Hct is hematocrit. E is the extraction fraction given by

$$E = 1 - e^{-\frac{PS}{F_p/V_t}} = 1 - e^{-\frac{PS}{F_b(1-Hct)/V_t}}, \tag{3}$$

where PS is the capillary permeability-surface area product per unit volume of tissue. Equation (3) is called "Renkin-Crone equation" [21, 22]. It should be noted that K^{trans} and k_{ep} are referred to as "K_1" and "k_2", respectively, in [23].

Solving Eq. (2) with the assumption that the initial conditions are zero yields

$$C_t(t) = K^{trans} \int_0^t C_p(\tau) e^{-k_{ep}(t-\tau)} d\tau + v_p C_p(t). \tag{4}$$

This equation has been used to analyze DCE-CT or DCE-MRI data in a number of studies [23].

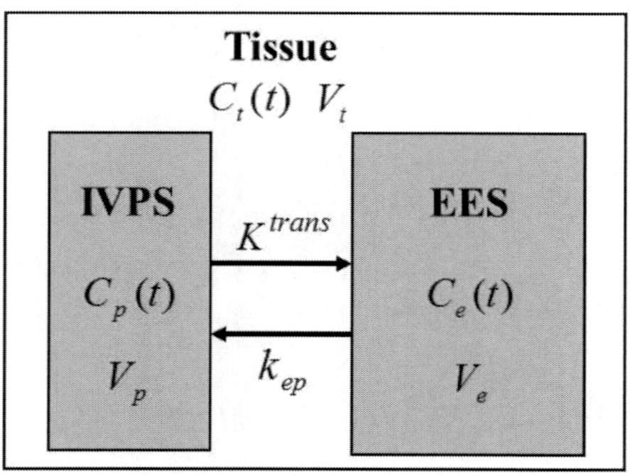

Figure 1. Two-compartment model consisting of an intravascular plasma space (IVPS) and extravascular-extracellular space (EES). K^{trans} is the volume transfer constant of contrast agent between the IVPS and EES and k_{ep} represents the transfer rate constant between the IVPS and EES. $C_t(t)$, $C_p(t)$, and $C_e(t)$ are the concentrations of contrast agent at time t in the tissue, IVPS, and EES, respectively. V_t, V_p, and V_e denote the volumes of the tissue, IVPS, and EES, respectively.

2.2.2. Three-Compartment Model

2.2.2.1. Laplace Transform

Figure 2 illustrates a mathematical model used to describe the kinetics of [^{18}F]fluoro-2-deoxy-D-glucose (FDG) (an analogue of glucose) transports in brain tissue and to determine the transport rate constants of FDG and to convert the radioactivity measurements to metabolic rates [24]. This model consists of three compartments, i.e., FDG in plasma, FDG in tissue, and FDG-6-PO_4 in tissue. We denote the concentrations of FDG or FDG-6-PO_4 in the three compartments at time t by $C_p(t)$, $C_E(t)$, and $C_M(t)$, respectively, as shown in Figure 2. The boundary between the plasma compartment and the tissue's FDG compartment represents the blood brain barrier for the transport of FDG. The boundary between the compartments of tissue's FDG and tissue's FDG-6-PO_4 is not a physical barrier. It represents the enzyme catalyzed chemical reactions between FDG and FDG-6-PO_4 [24].

In the model shown in Figure 2, the rates of change of FDG and FDG-6-PO_4 concentrations in tissue are equal to the net transportation of FDG and FDG-6-PO_4 into their compartments. That is:

$$\begin{cases} \frac{dC_E(t)}{dt} = K_1 C_p(t) - (k_2 + k_3)C_E(t) + k_4 C_M(t) \\ \frac{dC_M(t)}{dt} = k_3 C_E(t) - k_4 C_M(t) \end{cases}, \tag{5}$$

where $C_E(t)$ and $C_M(t)$ denote the concentrations of FDG and FDG-6-PO_4 in tissue at time t, respectively. $C_p(t)$ denotes the concentration of FDG in plasma. K_1, k_2, k_3, and k_4 denote the transfer rate constants between compartments. Applying the Laplace transform to Eq. (5) yields:

$$\begin{cases} sC_E(s) = K_1 C_p(s) - (k_2 + k_3)C_E(s) + k_4 C_M(s) \\ sC_M(s) = k_3 C_E(s) - k_4 C_M(s) \end{cases}, \tag{6}$$

where $C_E(s)$ and $C_M(s)$ denote the Laplace transforms of $C_E(t)$ and $C_M(t)$, respectively. s is a parameter for the Laplace transform. In this case, initial conditions were assumed to be zero, i.e., $C_E(0) = C_M(0) = 0$. From Eq. (6), $C_E(s)$ and $C_M(s)$ can be expressed in terms of $C_p(s)$ as:

$$\begin{cases} C_E(s) = \frac{K_1}{\alpha_2-\alpha_1}\left(\frac{k_4-\alpha_1}{s+\alpha_1} - \frac{k_4-\alpha_2}{s+\alpha_2}\right) C_p(s) \\ C_M(s) = \frac{K_1 k_3}{\alpha_2-\alpha_1}\left(\frac{1}{s+\alpha_1} - \frac{1}{s+\alpha_2}\right) C_p(s) \end{cases}, \quad (7)$$

where

$$\begin{cases} \alpha_1 = \frac{k_2+k_3+k_4-\sqrt{(k_2+k_3+k_4)^2-4k_2k_4}}{2} \\ \alpha_2 = \frac{k_2+k_3+k_4+\sqrt{(k_2+k_3+k_4)^2-4k_2k_4}}{2} \end{cases}. \quad (8)$$

If we use the properties of the unilateral Laplace transform and inverse Laplace transform, we obtain:

$$\begin{cases} C_E(t) = \frac{K_1}{\alpha_2-\alpha_1}[(k_4-\alpha_1)e^{-\alpha_1 t} - (k_4-\alpha_2)e^{-\alpha_2 t}]\otimes C_p(t) \\ C_M(t) = \frac{K_1 k_3}{\alpha_2-\alpha_1}(e^{-\alpha_1 t} - e^{-\alpha_2 t})\otimes C_p(t) \end{cases}, \quad (9)$$

where ⊗ denotes the operation of convolution integral.

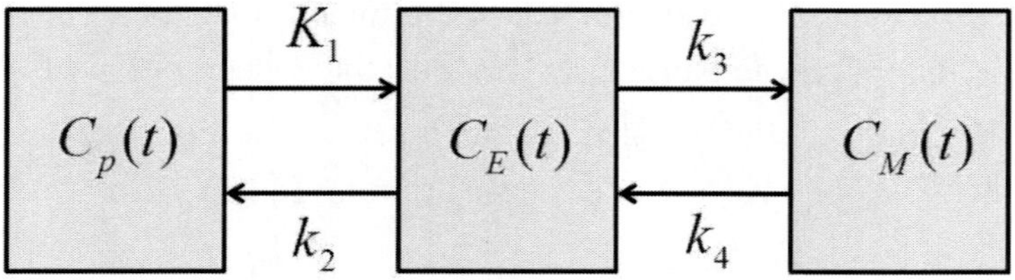

Figure 2. Three-compartment model used to describe the behavior of [^{18}F]fluoro-2-deoxy-D-glucose (FDG) in brain tissue. $C_E(t)$ and $C_M(t)$ denote the concentrations of FDG and FDG-6-PO_4 in brain tissue at time *t*, respectively. $C_p(t)$ denotes the concentration of FDG in plasma at time *t*. K_1, k_2, k_3, and k_4 denote the transfer rate constants between compartments.

2.2.2.2. Matrix Operation

Expressing simultaneous differential equations given by Eq. (5) in a matrix form yields:

$$\frac{d\mathbf{C}}{dt} = \mathbf{AC} + \mathbf{B}, \tag{10}$$

where

$$\mathbf{A} = \begin{bmatrix} -(k_2 + k_3) & k_4 \\ k_3 & -k_4 \end{bmatrix}, \tag{11}$$

$$\mathbf{B} = \begin{bmatrix} K_1 C_{\mathrm{p}}(t) \\ 0 \end{bmatrix}, \tag{12}$$

and

$$\mathbf{C} = \begin{bmatrix} C_{\mathrm{E}}(t) \\ C_{\mathrm{M}}(t) \end{bmatrix}. \tag{13}$$

The general solution of Eq. (10) is given by:

$$\mathbf{C}(t) = \int_0^t e^{(t-\tau)\mathbf{A}}\mathbf{B}(\tau)d\tau + e^{t\mathbf{A}}\mathbf{C}(0), \tag{14}$$

where $\mathbf{C}(0)$ denotes the initial conditions of $\mathbf{C}(t)$, i.e., $\mathbf{C}(0) = [C_{\mathrm{E}}(0) \quad C_{\mathrm{M}}(0)]^T$. T denotes the matrix transpose. $e^{(t-\tau)\mathbf{A}}$ and $e^{t\mathbf{A}}$ in Eq. (14) denote the matrix exponential. If **A** is diagonalizable, $e^{t\mathbf{A}}$ can be given by [25]

$$e^{t\mathbf{A}} = \mathbf{V}\begin{bmatrix} e^{\lambda_1 t} & 0 \\ 0 & e^{\lambda_2 t} \end{bmatrix}\mathbf{V}^{-1}, \tag{15}$$

where λ_1 and λ_2 denote the eigenvalues of **A**, and **V** is the eigenvector matrix. $\mathbf{V}^{-1}$ is the inverse matrix of **V**. The eigenvalues of **A** given by Eq. (11) can be obtained by solving the following matrix equation:

$$\det(\mathbf{A} - \lambda\mathbf{E}) = \begin{vmatrix} -(k_2 + k_3) - \lambda & k_4 \\ k_3 & -k_4 - \lambda \end{vmatrix} = \lambda^2 + (k_2 + k_3 + k_4)\lambda + k_2 k_4 = 0 \tag{16}$$

where **E** is the 2-by-2 identity matrix. Solving Eq. (16) yields

$$\lambda_1 = -\alpha_1 \tag{17}$$

and

$$\lambda_2 = -\alpha_2, \tag{18}$$

where α_1 and α_2 are given by Eq. (8). The eigenvector $\mathbf{V}_\lambda$ for an eigenvalue λ can be obtained by solving the following equation:

$$\mathbf{A}\mathbf{V}_\lambda = \lambda\mathbf{V}_\lambda. \tag{19}$$

Solving Eq. (19) for $\lambda = \lambda_1$ and $\lambda = \lambda_2$ yields the following equation for **V**

$$\mathbf{V} = \begin{bmatrix} \frac{k_4+\lambda_1}{k_3} & \frac{k_4+\lambda_2}{k_3} \\ 1 & 1 \end{bmatrix}. \tag{20}$$

Note that one element of the eigenvector was assumed to be unity in solving Eq. (19). The inverse matrix of **V** ($\mathbf{V}^{-1}$) can be obtained from $\mathbf{V}\mathbf{V}^{-1} = \mathbf{E}$ as:

$$\mathbf{V}^{-1} = \begin{bmatrix} \frac{k_3}{\lambda_1-\lambda_2} & -\frac{k_4+\lambda_2}{\lambda_1-\lambda_2} \\ -\frac{k_3}{\lambda_1-\lambda_2} & \frac{k_4+\lambda_1}{\lambda_1-\lambda_2} \end{bmatrix}. \tag{21}$$

Substituting Eqs. (20) and (21) into Eq. (14) yields the following equation as the solution of Eq. (10):

$$\mathbf{C}(t) = K_1 \begin{bmatrix} \frac{k_4+\lambda_1}{\lambda_1-\lambda_2} e^{\lambda_1 t} \otimes C_\mathrm{p}(t) - \frac{k_4+\lambda_2}{\lambda_1-\lambda_2} e^{\lambda_2 t} \otimes C_\mathrm{p}(t) \\ \frac{k_3}{\lambda_1-\lambda_2} e^{\lambda_1 t} \otimes C_\mathrm{p}(t) - \frac{k_3}{\lambda_1-\lambda_2} e^{\lambda_2 t} \otimes C_\mathrm{p}(t) \end{bmatrix} + e^{t\mathbf{A}}\mathbf{C}(0). \quad (22)$$

Note that we used Eq. (15) in deriving Eq. (22). When $\mathbf{C}(0) = \mathbf{0}$ in Eq. (22), Eq. (22) becomes equal to the solution obtained by the Laplace transform (Eq. (9)).

Figure 3 shows the time courses of the sum of FDG and FDG-6-PO_4 concentrations in tissue, i.e., $C_\mathrm{E}(t) + C_\mathrm{M}(t)$ obtained by the Laplace transform (solid line) and matrix operation (open circles), together with $C_\mathrm{p}(t)$ (closed circles). In these cases, K_1, k_2, k_3, and k_4 were assumed to be 0.120 mL/min/g, 0.107 min^{-1}, 0.044 min^{-1}, and 0.0059 min^{-1}, respectively. As shown in Figure 3, the solutions obtained by the Laplace transform and matrix operation agreed completely.

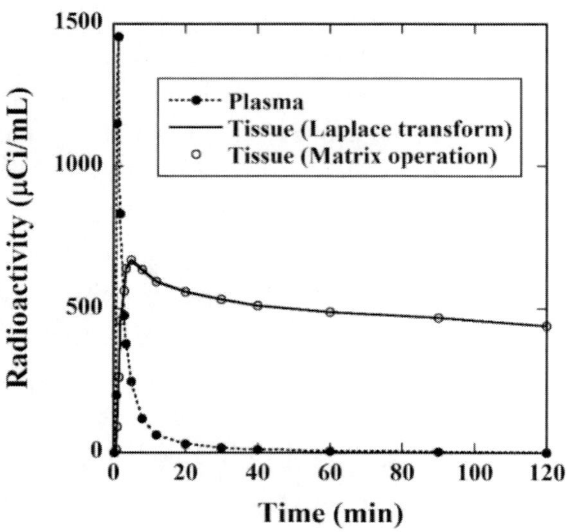

Figure 3. Time-activity curves of FDG and FDG-6-PO_4 in tissue obtained using the three-compartment model illustrated in Figure 2. The solid line and open circles show cases when using the Laplace transform and matrix operation, respectively. The closed circles show the time-activity curve of FDG in plasma.

2.2.3. Dual-Input Single-Compartment Model

Quantitative measurement of hepatic perfusion has the potential to provide important information in the assessment and management of various liver diseases and in the determination of their outcome [26, 27]. The non-invasive measurement of regional blood flow within the liver is of particular clinical interest, since changes in tumor blood flow have been recognized as an important factor for assessing the treatment response of hepatic cancers [28, 29]. The utility of hepatic perfusion characterization relies on the resolution of each component of its dual blood supply, i.e., the hepatic artery and the portal vein, because contributions from each are altered in many diseases [30].

While the two-compartment model shown in Figure 1 is applicable to the study of blood flow in various organs, it must be modified to reflect the unique delivery environment of the liver, where the tissue is fed by both arterial and portal inputs [29]. Two methods to accomplish this include using two separate input functions or using a combined input function. The latter method has been used previously such that a flow-weighted sum of both input functions was applied to hepatic FDG uptake, assuming a constant arterial-to-portal ratio across all regions of the liver [31]. However, such a fixed ratio system does not allow for the measurement of independent regional arterial and portal flow fluctuations. A more flexible approach is to incorporate the two separate input functions into the single-compartment model (dual-input single-compartment model), whereby the second input function reflects the tracer supplied by the portal vein. In this case, the model complexity increases and a third parameter for the local portal venous blood flow has to be estimated from the data [29].

In the early 1970s, it was discovered through necropsy that lung tumors also have a dual vascular supply [3], indicating that lung tumors can also become the target for tracer kinetic analysis using a dual-input single-compartment model. Recent advances in X-ray CT technology, particularly the development of a CT scanner with a large detector system composed of 320-detector row enabled one temporally uniform acquisition of the pulmonary artery, the aorta, the left atrium, and the lesion studied, which makes it possible to quantify the dual blood supply (pulmonary artery and bronchial artery) in lung cancer. The quantification of the dual blood supply to lung tumors and the relative portions of each can potentially aid in the management of lung cancer [3].

The compartment model used to describe the kinetic behavior of the contrast agent in the liver is illustrated in Figure 4. This model consists of a single compartment with two inputs. When using this model, the differential equation describing the kinetic behavior of contrast agent in the liver region of interest is given by

$$\frac{dC_L(t)}{dt} = K_{1a}C_{ha}(t)+K_{1p}C_{pv}(t) - k_2C_L(t), \tag{23}$$

where $C_L(t)$, $C_{ha}(t)$, and $C_{pv}(t)$ are the concentrations of the contrast agent at time t in the liver region of interest, hepatic artery or aorta, and portal vein, respectively. K_{1a}, K_{1p}, and k_2 are the rate constants for the exchanges of contrast agent from the hepatic artery or aorta to the liver, from the portal vein to the liver, and from the liver to plasma, respectively. It should be noted that whole-blood perfusion is obtained by dividing plasma perfusion by 1 minus the small-vessel hematocrit, as in Eq. (3). In deriving Eq. (23), the transit times of the contrast agent from the hepatic artery (alternatively aorta) and portal vein regions of interest to the liver region of interest were assumed to be negligible for simplicity. When these transit times are not negligible, they can be taken into account by replacing $C_{ha}(t)$ and $C_{pv}(t)$ in Eq. (23) by $C_{ha}(t-\tau_a)$ and $C_{pv}(t-\tau_p)$, respectively, where τ_a and τ_p denote the transit time of the contrast agent from the hepatic artery or aorta and portal vein to the liver, respectively [32]. Note that when analyzing the kinetic behavior of the contrast agent in lung tumors, $C_L(t)$, $C_{ha}(t)$, and $C_{pv}(t)$ in Eq. (23) correspond to the concentrations of the contrast agent in the lung tumor, pulmonary artery, and bronchial artery, respectively, and K_{1a}, K_{1p}, and k_2 correspond to the rate constants for the exchanges of contrast agent from the pulmonary artery to the lung tumor, from the bronchial artery to the lung tumor, and from the lung tumor to plasma, respectively.

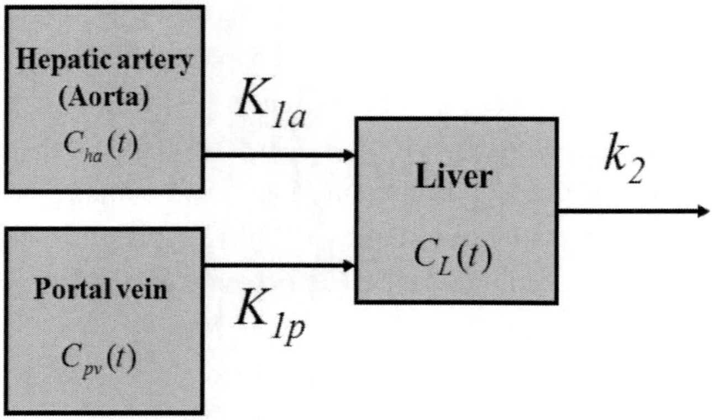

Figure 4. Dual-input single-compartment model used to describe the behavior of contrast agent in the liver. K_{1a}, K_{1p}, and k_2 are the rate constants for the exchanges of contrast agent from the hepatic artery or aorta to the liver, from the portal vein to the liver, and from the liver to plasma, respectively.

When the kinetic model includes a vascular term, Eq. (23) becomes:

$$\frac{d\{C_L(t)-v_0[C_{ha}(t)+C_{pv}(t)]\}}{dt} = K_{1a}C_{ha}(t)+K_{1p}C_{pv}(t) - k_2\{C_L(t) - v_0[C_{ha}(t) + C_{pv}(t)]\}, \quad (24)$$

where v_0 is the capillary plasma volume per unit volume of tissue. v_0 is given by a weighted average of v_{ha} and v_{pv} as [33]:

$$v_0 = \frac{v_{ha}C_{ha}(t)+v_{pv}C_{pv}(t)}{C_{ha}(t)+C_{pv}(t)}, \quad (25)$$

where v_{ha} and v_{pv} denote the relative volume fractions of the hepatic arterial plasma and the portal-venous plasma, respectively. Solving Eq. (24) with the assumption that the initial conditions are zero yields:

$$C_L(t) = \int_0^t [K_{1a}C_{ha}(\tau)+K_{1p}C_{pv}(\tau)]e^{-k_2(t-\tau)}d\tau + v_0[C_{ha}(t) + C_{pv}(t)]. \quad (26)$$

2.3. Model-Independent Method

With a model-independent method, kinetic parameters such as perfusion are quantified from the tissue response function obtained by deconvolution between the measured tissue time-concentration curve and the vascular input function.

The relationship between the time-concentration curve of contrast agent in tissue ($C_t(t)$) and that in a feeding artery, i.e., the arterial input function ($C_a(t)$) can be described as convolution integral:

$$C_t(t) = \frac{F}{V_t}\int_0^t C_a(\tau)R(t-\tau)d\tau = C_a(t)\otimes R_t(t), \tag{27}$$

where F and V_{t} denote blood flow and the volume of tissue, respectively. In Eq. (27), $R_t(t)$ $(= FR(t)/V_{\mathrm{t}})$ is the tissue response function which is the amount of the contrast agent in the tissue in an idealized perfusion experiment, where a unit area bolus is instantaneously injected ($R(0) = 1$) and subsequently washed out by perfusion ($R(\infty) = 0$). The procedure of finding a solution for the tissue response function that satisfies Eq. (27) is called "deconvolution analysis". It is known from Eq. (27) that the initial height of the deconvolved tissue response function equals F/V_{t}, i.e., perfusion.

Because the model-independent method only requires minimal user interaction or expertise in modeling, this approach appears to be easier to implement than the model-dependent method. Thus, the model-independent method has been applied to the quantification of perfusion in various organs such as brain [34], heart [35], and lung [36]. Furthermore, this method has been applied to the evaluation of the hemodynamic change due to angiogenesis in tumors [37] and monitoring the effect of therapeutic strategies on tumor vascular activity [38].

2.3.1. Truncated Singular Value Decomposition

If it is assumed that the functions $C_a(t)$ and $R(t)$ in Eq. (27) are constant over small time intervals, Δt, Eq. (27) is reduced to:

$$C_t(t_j) = \frac{F}{V_t} \sum_{i=1}^{n} C_a(t_i) R(t_j - t_i)\, \Delta t, \tag{28}$$

where t_i and *n* denote the time of the *i*th frame and the total number of frames, respectively. It should be noted that it was assumed that $t_1 = 0$ and the duration of each frame (Δt) is equal when converting Eq. (27) to Eq. (28). Equation (28) can then be formulated as a matrix equation:

$$\mathbf{C} = \mathbf{A} \cdot \mathbf{B}, \tag{29}$$

where

$$\mathbf{A} = \begin{bmatrix} C_a(t_1) & 0 & \cdots & 0 \\ C_a(t_2) & C_a(t_1) & \cdots & 0 \\ \vdots & \vdots & \ddots & \vdots \\ C_a(t_n) & C_a(t_{n-1}) & \cdots & C_a(t_1) \end{bmatrix}, \tag{30}$$

$$\mathbf{B} = \frac{F}{V_t} \begin{bmatrix} R(t_1) \\ R(t_2) \\ \vdots \\ R(t_n) \end{bmatrix} \Delta t, \tag{31}$$

and

$$\mathbf{C} = \begin{bmatrix} C_t(t_1) \\ C_t(t_2) \\ \vdots \\ C_t(t_n) \end{bmatrix}. \tag{32}$$

To solve Eq. (29) for **B** is called "deconvolution", which is an inverse problem and is a well-known ill-posed problem especially when there exists statistical noise. Thus, it is necessary to convert an ill-posed problem to a well-posed one to solve Eq. (29). The procedure of modifying an ill-posed problem to obtain a robust solution is called "regularization". Of the regularization methods, the truncated singular value decomposition is a popular method for obtaining regularized solutions for ill-posed inverse problems [14, 39].

With singular value decomposition, the matrix **A** in Eq. (29) can be expressed as the product of an $n \times n$ column-orthogonal matrix **U**, an $n \times n$ diagonal matrix **W** and the transpose of an $n \times n$ orthogonal matrix **V** [17], i.e.,

$$\mathbf{A} = \mathbf{UWV}^T = \mathbf{U}[\mathrm{diag}(w_i)]\mathbf{V}^T, \tag{33}$$

where w_i $(i = 1,2,\cdots,n)$ are the diagonal elements of **W** (singular values) which are nonnegative and can be ordered such that:

$$w_1 \geq w_2 \geq \cdots \geq w_n \geq 0. \tag{34}$$

The singular value decomposition constructs the matrices so that the inverse of **A** can be written as:

$$\mathbf{A}^{-1} = \mathbf{V}[\mathrm{diag}(1/w_i)]\mathbf{U}^T, \tag{35}$$

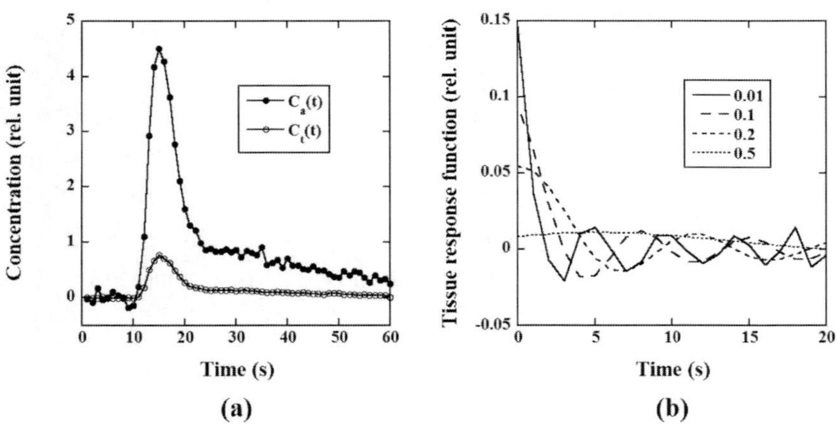

Figure 5. Example of the application of truncated singular value decomposition. (a) Time-concentration curves in a feeding artery ($C_a(t)$) and tissue ($C_t(t)$). (b) Tissue response functions obtained by deconvolution using the truncated singular value decomposition with various threshold values (0.01, 0.1, 0.2, and 0.5).

where $\mathbf{U}^T$ is the transpose of $\mathbf{U}$. The matrix $\mathbf{B}$ can then be calculated as:

$$\mathbf{B} = \mathbf{V}[\mathrm{diag}(1/w_i)](\mathbf{U}^T\mathbf{c}), \tag{36}$$

which contains the elements of $R(t)$.

With truncated singular value decomposition, if w_i was smaller than the maximal value of w_i multiplied by a threshold value, $1/w_i$ in Eq. (36) was replaced by zero. This threshold value can be thought of as a regularization parameter. The F/V_t value is obtained from the initial height (or maximum value) of $\mathbf{B}$ calculated from Eq. (36) [14, 39].

Figure 5 shows an example of the results obtained using the truncated singular value decomposition. Figure 5(b) shows an example of the tissue response functions obtained from $C_a(t)$ and $C_t(t)$ shown in Figure 5(a) by deconvolution using the truncated singular value decomposition with various threshold values (0.01, 0.1, 0.2, and 0.5).

As shown in Figure 5, the tissue response function obtained by deconvolution using the truncated singular value decomposition depends largely on the threshold value. When the threshold value is small, the tissue response function oscillates largely. On the other hand, the oscillation of the tissue response function decreases with increasing threshold value, but the initial height of the tissue response function decreases. The choice of a suitable regularization parameter is generally important, but can be difficult without auxiliary information about the problem being solved. The threshold value used in the truncated singular value decomposition is usually determined empirically by inspection of the results. Several methods, however, have been proposed to determine the optimal threshold value for truncated singular value decomposition automatically. These methods include a method using generalized cross validation [40] and L-curve method [41].

2.3.2. Spectral Analysis

Spectral analysis is based on the formulation of a constrained linear optimization problem, and is a development of an approach originally applied by Tobler and Engel [42] to the analysis of data obtained *in vitro*.

Spectral analysis was first introduced by Cunningham and Jones [43] in the field of nuclear medicine as a technique for use with dynamic PET studies. This technique is based on the actual kinetic components which can be identified within a given datum set, rather than being restricted, for example, to a predefined compartment model, and allows for the distinction between transit components associated with tracer in the vasculature, and reversible and irreversible components involved in the tissue without *a prior* assumptions regarding the number of components [43]. This method provides a spectrum of the kinetic components which are involved in the regional uptake and partitioning of tracer from the blood to the tissue [43]. This technique allows the tissue impulse response function to be derived with minimal modeling assumptions [43].

In spectral analysis, the concentration of contrast agent or radioactivity in the tissue at a given time t ($C(t)$) is modeled as a convolution of the vascular input function ($C_a(t)$) with a sum of k exponential terms as:

$$C(t) = \sum_{i=1}^{k} \alpha_i \cdot \int_0^t C_a(\tau) e^{-\beta_i(t-\tau)} d\tau, \tag{37}$$

where α_i and β_i ($\beta_1 < \beta_2 < \cdots < \beta_k$) are assumed to be positive or zero. The upper limit, k, represents the maximum number of terms to be included in the model and this is set to a large number. When $C_a(t)$ is replaced by Dirac's delta function ($\delta(t)$) in Eq. (37), the tissue impulse response function ($IRF(t)$) can be given by:

$$IRF(t) = \sum_{i=1}^{k} \alpha_i \cdot e^{-\beta_i t}. \tag{38}$$

2.3.2.1. Calculation of Spectrum

When the β values are given beforehand, the convolution integrals in Eq. (37) can be calculated using the measured $C_a(t)$. Then, the discrete form of Eq. (37) can be expressed by the following linear algebraic equation:

$$\begin{cases} C(t_1) = \alpha_1 \cdot f_1(\beta_1) + \alpha_2 \cdot f_1(\beta_2) + \cdots + \alpha_k \cdot f_1(\beta_k) \\ C(t_2) = \alpha_1 \cdot f_2(\beta_1) + \alpha_2 \cdot f_2(\beta_2) + \cdots + \alpha_k \cdot f_2(\beta_k) \\ \vdots \\ C(t_m) = \alpha_1 \cdot f_m(\beta_1) + \alpha_2 \cdot f_m(\beta_2) + \cdots + \alpha_k \cdot f_m(\beta_k) \end{cases} . \tag{39}$$

where m denotes the time index and $f_i(\beta_j)$ is given by:

$$f_i(\beta_j) = \int_0^{t_i} C_a(\tau) e^{-\beta_j(t_i-\tau)} d\tau. \tag{40}$$

Equation (37) can be given in a matrix form as:

$$\mathbf{C} = \mathbf{A} \cdot \mathbf{B}, \tag{41}$$

where

$$\mathbf{A} = \begin{bmatrix} f_1(\beta_1) & f_1(\beta_2) & \cdots & f_1(\beta_k) \\ f_2(\beta_1) & f_2(\beta_2) & \cdots & f_2(\beta_k) \\ \vdots & \vdots & \vdots & \vdots \\ f_m(\beta_1) & f_m(\beta_2) & \cdots & f_m(\beta_k) \end{bmatrix}, \tag{42}$$

$$\mathbf{B} = \begin{bmatrix} \alpha_1 \\ \alpha_2 \\ \vdots \\ \alpha_k \end{bmatrix}, \tag{43}$$

and

$$\mathbf{C} = \begin{bmatrix} C(t_1) \\ C(t_2) \\ \vdots \\ C(t_m) \end{bmatrix}. \tag{44}$$

The α value corresponding to each β can be easily obtained from Eq. (41) using the non-negative least squares method [44]. We call these α values "spectrum" from an analogy of time-frequency analysis.

2.3.2.2. Derivation of Quantitative Parameters

When there is no intravascular component, the sum of the heights of the peaks appearing in the spectrum, that is, the α values, is equal to K_1. Therefore, K_1 is given by:

$$K_1 = IRF(0) = \alpha_1 + \alpha_2 + \cdots + \alpha_k. \tag{45}$$

The mean transit time (MTT) of tracer in the tissue is given by:

$$MTT = \frac{\int_0^\infty IRF(t)dt}{IRF(0)}. \tag{46}$$

Substituting Eq. (38) into Eq. (46) yields:

$$MTT = \frac{\alpha_1/\beta_1 + \alpha_2/\beta_2 + \cdots + \alpha_k/\beta_k}{\alpha_1 + \alpha_2 + \cdots + \alpha_k}. \tag{47}$$

2.3.2.3. Clinical Applications

Figure 6 shows an example of the results of spectral analysis when applied to the three-compartment model with 4 parameters (Figure 2), in which the dephosphorylation of FDG-6-PO$_4$ to FDG (k_4) exists. Figure 6(c) shows the spectrum calculated from Eq. (41). As shown in Figure 6(c), there were two peaks in the spectrum. One peak was located at $\beta \approx \alpha_1$ (0.0049 min^{-1}) calculated from Eq. (8) and the other peak was located at $\beta \approx \alpha_2$ (0.0954 min^{-1}) calculated from Eq. (8). Figure 6(d) shows $IRF(t)$ calculated from Eq. (38). As shown in Figure 6(d), $IRF(t)$ decreased with time and $IRF(0)$ was 0.076 mL/min/g, which is nearly equal to the assumed value of K_1 (0.077 mL/min/g).

In addition to the above example, spectral analysis can be applied to various dynamic data acquired by planar scintigraphy, SPECT or PET [45]. For example, Fukui et al. [46] applied spectral analysis to hepatobiliary dynamic scintigraphy with ^{99m}Tc-*N*-pyridoxyl-5-methyltryptophan (PMT) for assessment of liver function in chronic liver diseases and regional function of irradiated liver, and reported that this method helps clarify changes in regional function of the irradiated liver. We applied spectral analysis to the renal uptake rate measurement of ^{99m}Tc-dimercaptosuccinic acid (DMSA) and found that spectral analysis is useful for the quantification of the renal uptake rate of DMSA [47]. We also applied spectral analysis to the estimation of the brain perfusion index (BPI) for measurement of cerebral blood flow (CBF) using ^{99m}Tc compounds [48], and reported that spectral analysis can provide a more reliable BPI for quantifying CBF than the conventional method using graphical analysis.

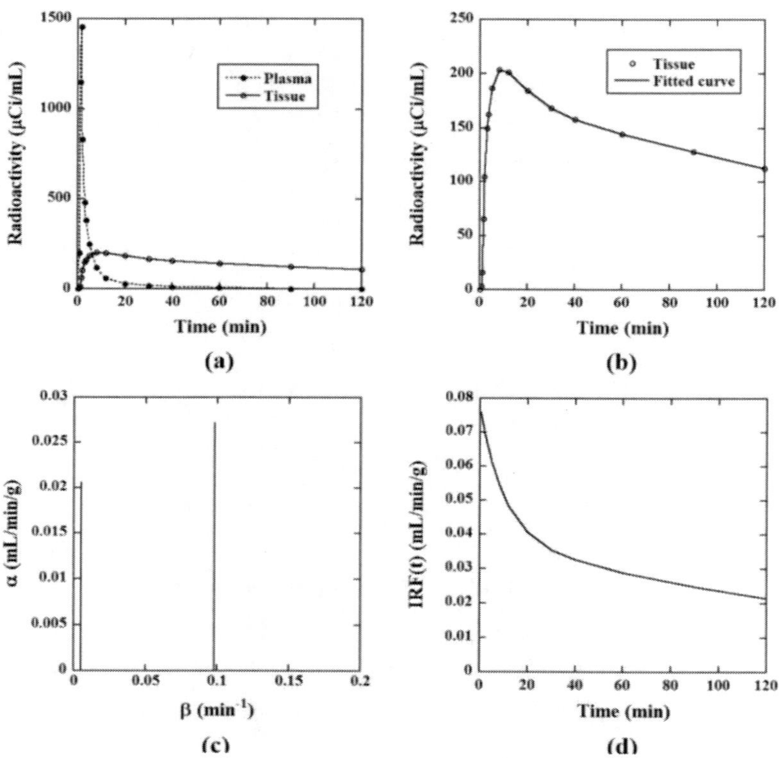

Figure 6. Example of spectral analysis applied to the three-compartment model illustrated in Figure 2. The closed and open circles in (a) show the time-activity curves of FDG in plasma and tissue, respectively. The open circles and solid line in (b) show the measured time-activity curve in tissue and that estimated from the tissue impulse response function ($IRF(t)$) obtained by spectral analysis, respectively. (c) and (d) show the spectrum and $IRF(t)$ obtained by spectral analysis, respectively. In these cases, K_1, k_2, k_3, and k_4 were assumed to be 0.077 mL/min/g, 0.050 min^{-1}, 0.041 min^{-1}, and 0.0093 min^{-1}, respectively.

Spectral analysis makes no *a priori* assumptions regarding the number of compartments or components required to describe the time course of tracer in the tissue. Spectral analysis can be applied to various dynamic data acquired by planar scintigraphy, SPECT or PET as an alternative approach to compartment analysis [45]. This analysis appears to be clinically useful, because it not only facilitates the interpretation of dynamic scintigraphic, SPECT or PET data, but also simplifies comparisons between regions and between subjects [45].

3. METHODS FOR KINETIC PARAMETER ESTIMATION

3.1. Nonlinear Least-Squares Method

In the case of a two-compartment model (Figure 1), fitting Eq. (4) to the time-concentration curve in the tissue obtained by DCE-CT allows the estimation of K^{trans}, k_{ep}, and v_p. In the case of a dual-input single-compartment model (Figure 4), fitting Eq. (26) to the time-concentration curve in the liver allows the estimation of K_{1a}, K_{1p}, k_2, and v_0. In general, these estimations can be performed by minimizing the sum of the squares of the differences between the data points and the model-estimated points, i.e., a merit function, using the nonlinear least-squares (NLSQ) method. For minimization of the merit function, there are several methods such as the Levenberg-Marquardt method [49] and the simplex method [50]. They are classified into two categories [51]. One is the method which requires both function evaluations and derivatives, and the other is the method which requires only function evaluations, not derivatives [51]. The Levenberg-Marquardt method belongs to the former category, whereas the simplex method belongs to the latter category. Algorithms using the derivative are somewhat more powerful than those using the only function, but not always enough so as to compensate for the additional calculations of derivatives [51].

The above methods search for the best combination of unknown parameters in a hypothetical model that predicts the measured data. These methods, however, are prone to estimating erroneous values. This is probably due to the noise intrinsic in the data and the inability of any search algorithm to find a global minimum of the merit function, i.e., the algorithm falls into a localized minimum of the merit function [49]. A strategy to avoid the selection of the localized minimum is to repeat the nonlinear least-squares fitting for a variety of initial guesses and to select the solution with the smallest value of the merit function.

When using a model-dependent method, it is often unclear which combination of unknown parameters best approximates the measured data. A commonly-used means to decide which model is the best is the corrected Akaike information criterion (cAIC) [52]. This criterion penalizes the goodness of the fit, based on not only the merit function (*SSR*) but also the number of estimated parameters (*M*) and sample size (*N*) [52]. That is, $cAIC$ is given by

$$cAIC = N \cdot \ln\left(\frac{SSR}{N}\right) + \frac{2N(M+1)}{N-M-2}. \tag{48}$$

The value of the $cAIC$ itself for a given data set has no meaning, and it can only be interpreted comparing the $cAIC$ of a series of models based on the same data set. Then, the model with the lowest $cAIC$ is the best model among all considered models [52].

Although the NLSQ method is a powerful tool for estimating kinetic parameters, it is not always suitable for generating parametric images of physiological parameters, because it is not efficient in terms of computation time. Furthermore, appropriate choices of initial guesses of the parameters are important in the NLSQ method; poor initial guesses result in long computation times or occasionally failure to converge on a good fit. Thus, to generate parametric images of physiological parameters using DCE-CT, an efficient method for estimating kinetic parameters is necessary, which is computationally efficient and does not need initial guesses of the parameters.

3.2. Linear Least-Squares Method

3.2.1. Two-Compartment Model

Integrating both sides of Eq. (2) with the assumption that the initial conditions are zero, and expressing the result in a discrete form yield [23]

$$C_t(t_k) = \left(K^{trans} + k_{ep}v_p\right)\int_0^{t_k} C_p(\tau)d\tau - k_{ep}\int_0^{t_k} C_t(\tau)d\tau + v_pC_p(t_k)\ (k = 1,2,\cdots,n). \tag{49}$$

Equation (49) can be given in a matrix form as

$$\mathbf{C} = \mathbf{A}\cdot\mathbf{B}, \tag{50}$$

where

$$\mathbf{A} = \begin{bmatrix} \int_0^{t_1} C_p(\tau)d\tau & -\int_0^{t_1} C_t(\tau)d\tau & C_p(t_1) \\ \int_0^{t_2} C_p(\tau)d\tau & -\int_0^{t_2} C_t(\tau)d\tau & C_p(t_2) \\ \vdots & \vdots & \vdots \\ \int_0^{t_n} C_p(\tau)d\tau & -\int_0^{t_n} C_t(\tau)d\tau & C_p(t_n) \end{bmatrix}, \tag{51}$$

$$\mathbf{B} = \begin{bmatrix} K^{trans} + k_{ep}v_p \\ k_{ep} \\ v_p \end{bmatrix}, \tag{52}$$

and

$$\mathbf{C} = \begin{bmatrix} C_t(t_1) \\ C_t(t_2) \\ \vdots \\ C_t(t_n) \end{bmatrix}. \tag{53}$$

When $C_t(t_k)$ and $C_p(t_k)$ $(k = 1,2,\cdots,n)$ are measured, Eq. (50) can be easily solved for the elements of $\mathbf{B}$ using the conventional linear least-squares (LLSQ) method [51], and finally K^{trans}, k_{ep}, and v_p can be easily obtained from the elements [23]. We call this method the "LLSQ method". The elements of **A** can be generated by numerical integration using trapezoidal rule [51].

It has been demonstrated that the LLSQ method can estimate the kinetic parameters faster and more accurately than the NLSQ method especially at low signal-to-noise ratio [23]. Furthermore, the LLSQ method does not need initial guesses of the parameters unlike the NLSQ method [23]. These results suggest that the LLSQ method is useful for kinetic analysis of contrast agent in the tissue using DCE-CT or DCE-MRI and especially for generating the parametric images of kinetic parameters by applying it pixel by pixel.

3.2.2. Three-Compartment Model

When $k_4 = 0$ in Eqs. (8) and (9), rearranging and expressing Eq. (9) in a discrete form yield [53]

$$C_t(t_k) = C_E(t_k) + C_M(t_k) = K_1 \int_0^{t_k} C_p(\tau)d\tau + K_1 k_3 \int_0^{t_k}\int_0^{t} C_p(\tau)\, d\tau dt - (k_2 + k_3)\int_0^{t_k} C_t(\tau)d\tau \ (k = 1,2,\cdots,n). \tag{54}$$

Equation (54) can be given in a matrix form as

$$\mathbf{C} = \mathbf{A} \cdot \mathbf{B}, \tag{55}$$

where

$$\mathbf{A} = \begin{bmatrix} \int_0^{t_1} C_p(\tau)d\tau & \int_0^{t_1}\int_0^{t} C_p(\tau)\, d\tau dt & -\int_0^{t_1} C_t(\tau)d\tau \\ \int_0^{t_2} C_p(\tau)d\tau & \int_0^{t_2}\int_0^{t} C_p(\tau)\, d\tau dt & -\int_0^{t_2} C_t(\tau)d\tau \\ \vdots & \vdots & \vdots \\ \int_0^{t_n} C_p(\tau)d\tau & \int_0^{t_n}\int_0^{t} C_p(\tau)\, d\tau dt & -\int_0^{t_n} C_t(\tau)d\tau \end{bmatrix}, \tag{56}$$

$$\mathbf{B} = \begin{bmatrix} K_1 \\ K_1 k_3 \\ k_2 + k_3 \end{bmatrix}, \tag{57}$$

and

$$\mathbf{C} = \begin{bmatrix} C_t(t_1) \\ C_t(t_2) \\ \vdots \\ C_t(t_n) \end{bmatrix}. \tag{58}$$

As in the two-compartment model (Figure 1), when $C_t(t_k)$ and $C_p(t_k)$ ($k = 1,2,\cdots,n$) are measured, Eq. (55) can be easily solved for the elements of **B** using the LLSQ method [51], and finally K_1, k_2, and k_3 can be easily obtained from the elements [53].

3.2.3. Dual-Input Single-Compartment Model

Integrating both sides of Eq. (24) with the assumption that the initial conditions are zero, and expressing the result in a discrete form yield [33]:

$$C_L(t_k) = (K_{1a} + k_2 v_0) \int_0^{t_k} C_{ha}(\tau)d\tau + \left(K_{1p} + k_2 v_0\right) \int_0^{t_k} C_{pv}(\tau)d\tau - k_2 \int_0^{t_k} C_L(\tau)d\tau + v_0\left[C_{ha}(t_k) + C_{pv}(t_k)\right]\ (k = 1,2,\cdots,n). \tag{59}$$

Equation (59) can be given in a matrix form as

$$\mathbf{C} = \mathbf{A} \cdot \mathbf{B}, \tag{60}$$

where

$$\mathbf{A} = \begin{bmatrix} \int_0^{t_1} C_{ha}(\tau)d\tau & \int_0^{t_1} C_{pv}(\tau)d\tau & -\int_0^{t_1} C_L(\tau)d\tau & C_{ha}(t_1) + C_{pv}(t_1) \\ \int_0^{t_2} C_{ha}(\tau)d\tau & \int_0^{t_2} C_{pv}(\tau)d\tau & -\int_0^{t_2} C_L(\tau)d\tau & C_{ha}(t_2) + C_{pv}(t_2) \\ \vdots & \vdots & \vdots & \vdots \\ \int_0^{t_n} C_{ha}(\tau)d\tau & \int_0^{t_n} C_{pv}(\tau)d\tau & -\int_0^{t_n} C_L(\tau)d\tau & C_{ha}(t_n) + C_{pv}(t_n) \end{bmatrix}, \tag{61}$$

$$\mathbf{B} = \begin{bmatrix} K_{1a} + k_2 v_0 \\ K_{1p} + k_2 v_0 \\ k_2 \\ v_0 \end{bmatrix}, \tag{62}$$

and

$$\mathbf{C} = \begin{bmatrix} C_L(t_1) \\ C_L(t_2) \\ \vdots \\ C_L(t_n) \end{bmatrix}. \tag{63}$$

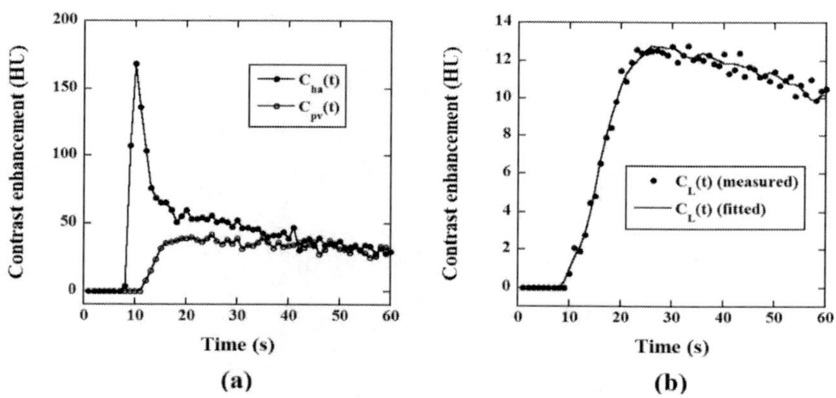

Figure 7. Example of the linear least-squares method applied to the dual-input single-compartment model illustrated in Figure 4. (a) Time-concentration curves in the hepatic artery ($C_{ha}(t)$) (closed circles) and portal vein ($C_{pv}(t)$) (open circles). (b) Measured (closed circles) and fitted time-concentration curves in the liver ($C_L(t)$) (solid line).

When $C_L(t_k)$, $C_{ha}(t_k)$, and $C_{pv}(t_k)$ $(k = 1,2,\cdots,n)$ are measured, Eq. (60) can be easily solved for the elements of **B** using the LLSQ method [51] as in Eq. (50), and finally K_{1a}, K_{1p}, k_2, and v_0 can be easily obtained from the elements [33].

Figure 7 shows an example of the LLSQ method for the dual-input single-compartment model applied to the DCE-CT data in the liver. As shown in Figure 7(b), the time-concentration curve in the liver estimated by the LLSQ method (solid line) is well fitted to that actually measured (closed circles).

It has been demonstrated by computer simulations that the LLSQ method can estimate the kinetic parameters in the liver with almost the same accuracy as the NLSQ method [33]. Furthermore, the LLSQ method can estimate them faster than the NLSQ method by a factor of approximately 10 [33]. These results suggest that the LLSQ method is useful for analyzing the kinetic behavior of contrast agent in the organ or tissue having a dual blood supply such as the liver and lung tumor, especially when generating the parametric images of kinetic parameters by applying it pixel by pixel.

3.3. Graphical Method

3.3.1. Two-Compartment Model

When using a two-compartment model illustrated in Figure 1, the differential equation describing the kinetic behavior of the contrast agent in the tissue is given by Eq. (4) [23]. Solving Eq. (4) for $C_t(t)$ with the assumption that the initial conditions are zero and k_{ep} is negligible during the sampling period yields

$$C_t(t) = K^{trans} \int_0^t C_p(\tau)d\tau + v_p C_p(t). \tag{64}$$

Dividing both sides of Eq. (64) by $C_p(t)$ yields

$$\frac{C_t(t)}{C_p(t)} = K^{trans} \frac{\int_0^t C_p(\tau)d\tau}{C_p(t)} + v_p. \tag{65}$$

K^{trans} and v_p can be estimated from Eq. (65) using the Gjedde-Patlak graphical method [54, 55]. In the Gjedde-Patlak graphical method, the normalized time given by $\int_0^t C_p(\tau)d\tau / C_p(t)$ is taken as the x axis and $C_t(t)/C_p(t)$ is taken as the y axis. K^{trans} and v_p can be obtained from the slope and y-intercept of the regression line fitted to the Gjedde-Patlak plot, respectively.

3.3.2. Three-Compartment Model

When $k_4 = 0$ in Eqs. (8) and (9), rearranging Eq. (9) yields

$$C_t(t) = C_\mathrm{E}(t) + C_\mathrm{M}(t) = \frac{K_1 k_3}{k_2+k_3}\int_0^t C_p(\tau)d\tau + \frac{k_2}{k_2+k_3}C_E(t). \quad (66)$$

Dividing both sides of Eq. (66) by $C_p(t)$ yields

$$\frac{C_t(t)}{C_p(t)} = \frac{K_1 k_3}{k_2+k_3}\frac{\int_0^t C_p(\tau)d\tau}{C_p(t)} + \frac{k_2}{k_2+k_3}\frac{C_E(t)}{C_p(t)}. \quad (67)$$

After some time, the reversible compartment concentration follows the plasma concentration so that $C_E(t)/C_p(t)$ is constant. As in Eq. (65), when the normalized time given by $\int_0^t C_p(\tau)d\tau/C_p(t)$ is taken as the x axis and $C_t(t)/C_p(t)$ is taken as the y axis, the tracer net uptake (K) given by $K_1k_3/(k_2+k_3)$ can be obtained from the slope of the regression line fitted to the Gjedde-Patlak plot.

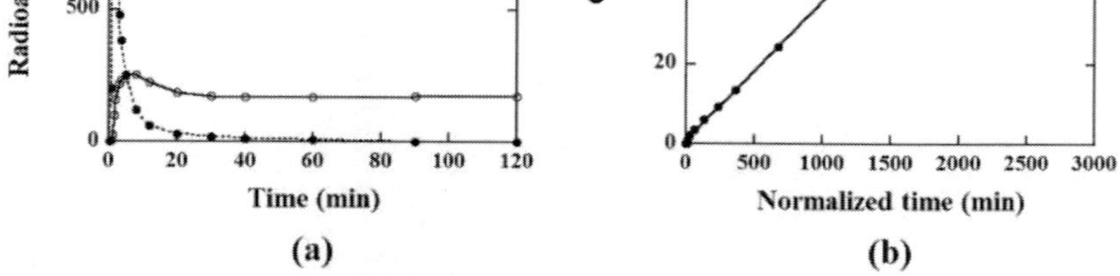

Figure 8. (a) Example of $C_p(t)$ (closed circles) and $C_t(t)$ (open circles) used for generating the Gjedde-Patlak plot. (b) Gjedde-Patlak plot, in which the normalized time given by $\int_0^t C_p(\tau)d\tau/C_p(t)$ is taken as the x axis and $C_t(t)/C_p(t)$ is taken as the y axis. The solid line represents the regression line, from which the tracer net uptake ($K = K_1k_3/(k_2+k_3)$) is obtained.

Figure 8(a) shows an example of $C_p(t)$ and $C_t(t)$ obtained by FDG PET, whereas Figure 8(b) shows the Gjedde-Patlak plot generated from these data. In Figure 8(b), the regression line fitted to the Gjedde-Patlak plot between 500 min and 2500 min is also shown by solid line, whose slope corresponds to K. In these cases, K_1, k_2, k_3, and k_4 were assumed to be 0.120 mL/min/g, 0.107 min^{-1}, 0.044 min^{-1}, and 0.0 min^{-1}, respectively. The slope of the regression line shown in Figure 8(b) was 0.0351 mL/min/g, which was equal to the true tracer net uptake (0.0350 mL/min/g).

The graphical method is superior to other methods in terms of the simplicity and easiness to implement, but attention should be paid to setting the range of linear regression for calculating the slope and y-intercept, because the resultant K^{trans}, v_p, and K values may vary depending on this range.

4. OTHERS

4.1. Adiabatic Approximation to the Tissue Homogeneity Model

When we take into account the effect of the leakage of contrast agent from the intravascular space to the extravascular space, it is necessary to use the distributed parameter model describing the concentration of contrast agent within the tissue as a function of both time and position. One of the models describing the tracer concentration in the vascular space as a function of both time and position along the capillary is the tissue homogeneity model [16], and St. Lawrence and Lee [17] derived an approximate closed-form solution to the tissue homogeneity model using the adiabatic approximation.

In the adiabatic approximation to the tissue homogeneity model, it is assumed that the impulse residue (retention) function ($R(t)$) is given by [17]

$$R(t) = \begin{cases} 1 & 0 \le t \le t_c \\ Ee^{-\frac{EF}{V_e}(t-t_c)} & t_c \le t \end{cases}, \tag{68}$$

where V_e is the distribution volume of contrast agent in the extravascular-extracellular space and t_c is the transit time through the capillary which is equal to the distribution volume of contrast agent in the intravascular space (V_i) divided by blood flow (F), i.e., $t_c = V_i/F$. E is the extraction fraction defined by Eq. (3). Inserting Eq. (68) into Eq. (27) yields [15]

$$C_t(t) = \frac{F}{V_t}\int_0^{t_c} C_a(t-\tau)d\tau + \frac{EF}{V_t}\int_{t_c}^{t} C_a(t-\tau)e^{-\frac{EF}{V_e}(t-t_c)}d\tau. \tag{69}$$

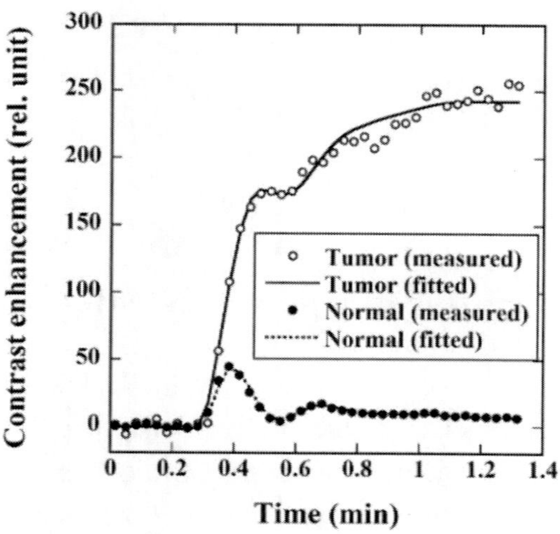

Figure 9. Example of the results obtained by applying the adiabatic approximation to the tissue homogeneity model to DCE-CT data acquired from a patient with meningioma. The open and closed circles show the measured data in a tumor and normal regions, respectively, whereas the solid and dotted lines show the fitted curves for a tumor and normal regions, respectively, using Eq. (69).

Figure 9 shows an example of the results obtained by applying the adiabatic approximation to the tissue homogeneity model to DCE-CT data acquired from a patient with meningioma. Six parameters (F/V_t, K_1 ($= EF$), PS, k_2 ($= EF/V_e$), t_c, and E) were estimated by fitting Eq. (69) to the DCE-CT data using the NLSQ method. E was calculated from F/V_t

and PS using Eq. (3). In Figure 9, open and closed circles show the measured data in a tumor and normal regions, respectively, whereas the solid and dotted lines show the fitted curves for a tumor and normal regions, respectively, using Eq. (69). In this case, F/V_t, K_1 ($= EF$), PS, k_2 ($= EF/V_e$), t_c, and E were obtained to be 82.6 mL/100mL/min, 54.1 mL/100mL/min, 88.0 mL/100mL/min, 2.2 min^{-1}, 8.4 s, and 0.7, respectively, for a tumor region, whereas they were 40.1 mL/100mL/min, 1.7 mL/100mL/min, 1.6 mL/100mL/min, 0.8 min^{-1}, 4.8 s, and 0.04, respectively, for a normal region, indicating that both the blood flow and permeability in the tumor are considerably increased as compared to those in the normal region. These results suggest that this approach is useful for evaluating perfusion and capillary permeability simultaneously and separately.

4.2. Empirical Mathematical Model

Recently, Fan et al. developed an empirical mathematic model (EMM) to describe contrast uptake and washout behavior without use of vascular input functions [56]. The EMM has been applied to DCE-MRI data to distinguish metastatic from nonmetastatic transplanted rodent prostate tumors [56], to distinguish between benign and malignant breast lesions [57], and to detect bowel inflammation in patients with Crohn's disease [58].

Various liver-specific contrast agents for MRI have been introduced to increase the accuracy of hepatic imaging [59]. One recently available hepatocyte-targeted contrast agent is gadolinium ethoxybenzyl diethylenetriamine pentaacetic acid (Gd-EOB-DTPA) [60]. Since Gd-EOB-DTPA partially accumulates in hepatocytes and bile via various transporters after intravenous injection, the signal intensity in the liver increases on T_1-weighted images [61]. The signal intensity of the liver after Gd-EOB-DTPA injection depends on the Gd-EOB-DTPA uptake by hepatocytes and bile excretion. It is known that the Gd-EOB-DTPA transport correlates with liver function. Therefore, the DCE-MRI with Gd-

EOB-DTPA appears to be a promising method for evaluating liver function.

The liver has a dual blood supply from the hepatic artery and the portal vein [33]. Thus, when analyzing the DCE-MRI data in the liver using a pharmacokinetic model, this unique environment of blood supply increases the model complexity. Furthermore, there are many cases in which it is difficult to measure these dual vascular inputs separately, especially in animal experiments. Thus, a simple and reliable method is desired for analyzing the DCE-MRI data, which does not need vascular input functions. Then, we tried to analyze the DCE-MRI data obtained with Gd-EOB-DTPA using the EMM [62].

First, we obtained the signal intensity at time t ($SI(t)$) by drawing the region of interest (ROI) on the liver parenchyma without vascular region with area of approximately 1000 pixels [62]. We estimated the contrast agent concentration as a function of time after contrast agent injection by comparing $SI(t)$ from a selected ROI to the signal intensity before contrast agent injection ($SI(0)$) as [56, 62]

$$RE(t) = \frac{SI(t)-SI(0)}{SI(0)} \times 100\ (\%). \tag{70}$$

For analysis, the EMM with 5 parameters [58] was used to fit the time-signal intensity curves obtained using DCE-MRI with Gd-EOB-DTPA:

$$RE(t) = \begin{cases} 0 & 0 \le t < t_0 \\ A \cdot \left[1 - e^{-\alpha(t-t_0)}\right]^q \cdot e^{-\beta(t-t_0)} & t_0 \le t \end{cases}, \tag{71}$$

where $RE(t)$ is the relative enhancement at time t, A the upper limit of the signal intensity, α the rate of contrast uptake (min^{-1}), β the rate of contrast washout (min^{-1}), q the parameter related to the slope of early uptake, and t_0 the rise time point (min). It should be noted that t_0 was added to the EMM with 4 parameters [58] as an additional parameter in this study. The above 5 parameters were calculated using the NLSQ method (simplex method) [50]. Appropriate choices of initial guesses of the parameters were important; poor initial guesses resulted in long computation times or occasionally failure to converge on a good fit. Based on our experiences, the initial guesses were taken to be as follows: A = the maximum value of relative enhancement, α=0.2 min^{-1}, β=0.03 min^{-1}, q=0.3, and t_0=2 min.

The area under the curve (AUC) was calculated by integrating Eq. (71) from t_0 to the last time point. The time to the maximum relative enhancement (T_{max}) was calculated by setting the derivative of Eq. (71) equal to zero:

$$T_{max} = \frac{1}{\alpha} \cdot \ln\left(1 + \frac{\alpha q}{\beta}\right). \tag{72}$$

The maximum relative enhancement (RE_{max}) was obtained by substituting Eq. (72) into Eq. (71) as:

$$RE_{max} = A \cdot [1 - e^{-\alpha T_{max}}]^q \cdot e^{-\beta T_{max}}. \tag{73}$$

The elimination half-life of relative enhancement ($T_{1/2}$) was also calculated by numerically solving the following equation using Newton-Raphson algorithm [51]:

$$A \cdot \left[1 - e^{-\alpha(T_{max} + T_{1/2})}\right]^q \cdot e^{-\beta(T_{max} + T_{1/2})} = \frac{RE_{max}}{2}. \tag{74}$$

In this study, we evaluated the concanavalin A (Con A)-induced liver injury in rats by analyzing the DCE-MRI data obtained with Gd-EOB-DTPA using the EMM mentioned above [62]. The rats were allocated into 3 groups (normal control (n = 6), Con A (10) (n = 6), and Con A (20) groups (n = 6)). The rats of the Con A (10) group were intravenously injected with Con A alone from the tail vein at a single dose of 10 mg/kg body weight (BW) dissolved in normal saline, whereas those of the Con A (20) group were intravenously injected with 20 mg/kg BW of Con A. In the normal control group, the rats were intravenously injected with the same volume of saline. The DCE-MRI studies were performed 24 hours after Con A or saline injection [62].

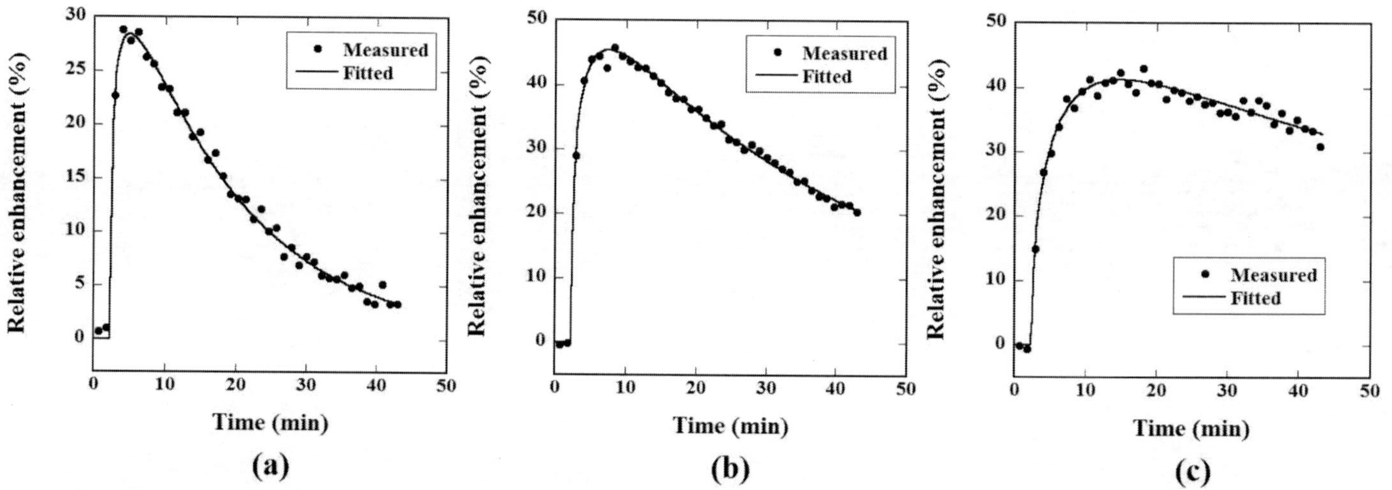

Figure 10. Examples of the results obtained by applying the EMM to DCE-MRI with Gd-EOB-DTPA in the normal control (a), Con A (10) (b), and Con A (20) groups (c). The closed circles and solid lines represent the measured and fitted data using Eq. (71), respectively.

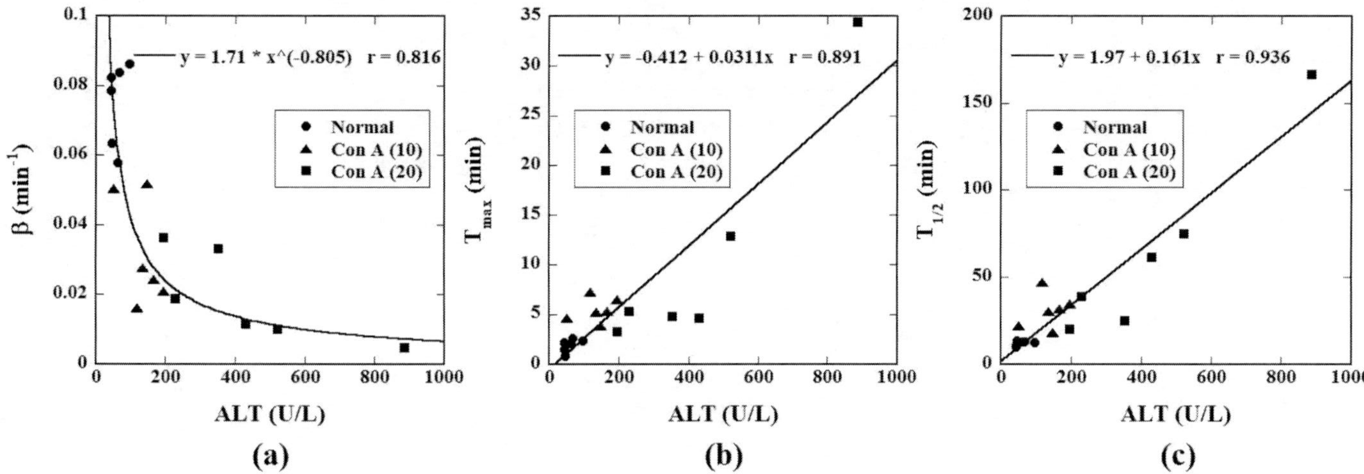

Figure 11. Correlations between β and serum alanine aminotransferase (ALT) (a), between T_{max} and ALT (b), and between $T_{1/2}$ and ALT (c).

Figure 10 shows examples of curve fitting using the EMM in the normal control (a), Con A (10) (b), and Con A (20) groups (c). As shown in Figure 10, a continuous increase and slow reduction of $RE(t)$ after peaking at T_{max} were induced in all groups, but signal profiles differed among the 3 groups. The EMM provided excellent fits to data from all rats despite large variations in their uptake and washout behavior, suggesting that the EMM is feasible for analyzing the DCE-MRI data in the Con A-induced liver injury. Our results demonstrated that, of the parameters extracted by the EMM approach, β, T_{max}, and $T_{1/2}$ correlated well with serum alanine aminotransferase (ALT) (biomarker of liver injury) levels (Figure 11), and are the sensitive and useful parameters for evaluating the Con A-induced liver injury.

As previously described, the EMM does not require vascular input functions to analyze the DCE-MRI data. Another advantage of the EMM approach is that this approach does not require making assumptions about underlying physiology or anatomy unlike other model-based approaches [20]. As pointed out by Fan et al. [57], however, the primary disadvantage of the EMM approach is that the parameters obtained by this approach do not correspond directly to identifiable physiological or anatomic features. However, this problem can be addressed by deriving equations that connect the parameters obtained by the EMM approach to physiological and anatomic parameters associated with various pharmacokinetic models as done by Fan et al. [57]. As reported by Hayton et al. [63], when an instantaneous injection of contrast agent was assumed, the equation describing the time-signal intensity curve in the two-compartment model of Brix et al. [7] becomes similar to that for the EMM.

Conclusion

In this chapter, we described the basics of mathematical modeling of tracer kinetics and various methods for quantifying kinetic parameters such as perfusion using PET, DCE-CT, or DCE-MRI, which are roughly classified into model-dependent and model-independent methods. We also described various methods for estimating the kinetic parameters, which include the NLSQ method, LLSQ method, and graphical method. Finally, we touched on the adiabatic approximation to the tissue homogeneity model and the EMM approach. We hope that these will be useful and helpful for readers to gain a deeper understanding of mathematical modeling of tracer kinetics for medical applications.

REFERENCES

[1] Boxenbaum, H. (1992). Pharmacokinetics: philosophy of modeling. *Drug Metabolism Reviews*, 24: 89–120.

[2] Phelps, M. E., Huang, S. C., Hoffman, E. J., Selin, C. E. and Kuhl, D. E. (1979). Tomographic measurement of local cerebral glucose metabolic rate in humans with (F-18) 2-fluoro-2-deoxy-D-glucose: validation of method. *Annuals of Neurology*, 6: 371–388.

[3] Yuan, X., Zhang, J., Ao, G., Quan, C., Tian, Y. and Li, H. (2012). Lung cancer perfusion: can we measure pulmonary and bronchial circulation simultaneously? *European Radiology*, 22: 1665–1671.

[4] Brix, G., Griebel, J., Kiessling, F. and Wenz, F. (2010). Tracer kinetic modelling of tumour angiogenesis based on dynamic contrast-enhanced CT and MRI measurements. *European Journal of Nuclear Medicine and Molecular Imaging*, 37: S30–S51.

[5] Morcos, S. K. and Thomsen, H. S. (2001). Adverse reactions to iodinated contrast media. *European Radiology*, 11: 1267–1275.

[6] Thakur, Y., McLaughlin, P. D. and Mayo, J. R. (2013). Strategies for radiation dose optimization. *Current Radiology Reports*, 1: 1–10.

[7] Brix, G., Semmler, W., Port, R., Schad, L. R., Layer, G. and Lorenz, W. J. (1991). Pharmacokinetic parameters in CNS Gd-DTPA enhanced MR imaging. *Journal of Computer Assisted Tomography*, 15: 621–628.

[8] Choyke, P. L., Dwyer, A. J. and Knopp, M. V. (2003). Functional tumor imaging with dynamic contrast-enhanced magnetic resonance imaging. *Journal of Magnetic Resonance Imaging*, 17: 509–520.

[9] Koh, T. S., Bisdas, S., Koh, D. M. and Thng, C. H. (2011). Fundamentals of tracer kinetics for dynamic contrast-enhanced MRI. *Journal of Magnetic Resonance Imaging*, 34: 1262–1276.

[10] Lee, T. Y. (2002). Functional CT: physiological models. *Trends in Biotechnology*, 20: S3–S10.

[11] Koh, T. S., Tan, C. K. M., Cheong, L. H. D. and Lim, C. C. T. (2006). Cerebral perfusion mapping using a robust and efficient method for deconvolution analysis of dynamic contrast-enhanced images. *Neuro Image*, 32: 643–653.

[12] Kershaw, L. E. and Buckley, D. L. (2006). Precision in measurements of perfusion and microvascular permeability with T_1-weighted dynamic contrast-enhanced MRI. *Magnetic Resonance in Medicine*, 56: 986–992.

[13] Meier, P. and Zierler, K. L. (1954). On the theory of the indicator-dilution method for measurement of blood flow and volume. *Journal of Applied Physiology*, 6: 731–744.

[14] Østergaard, L., Weisskoff, R. M., Chesler, D. A., Gyldensted, C. and Rosen, B. R. (1996). High resolution measurement of cerebral blood flow using intravascular tracer bolus passages. part 1: mathematical approach and statistical approach. *Magnetic Resonance in Medicine*, 36: 715–725.

[15] Murase, K. and Miyazaki, S. (2007). Error analysis of tumor blood flow measurement using dynamic contrast-enhanced data and model-independent deconvolution analysis. *Physics in Medicine and Biology*, 52: 2791–2805.

[16] Johnson, J. A. and Wilson, T. A. (1966). A model for capillary exchange. *American Journal of Physiology*, 210: 1299–1303.

[17] St. Lawrence, K. S. and Lee, T. Y. (1998). An adiabatic approximation to the tissue homogeneity model for water exchange in the brain. I. theoretical derivation. *Journal of Cerebral Blood Flow and Metabolism*, 18: 1365–1377.

[18] Tofts, P. S. (1997). Modeling tracer kinetics in dynamic Gd-DTPA MR imaging. *Journal of Magnetic Resonance Imaging*, 7: 91–101.

[19] Buckley, D. L. (2002). Uncertainty in the analysis of tracer kinetics using dynamic contrast-enhanced T_1-weighted MRI. *Magnetic Resonance in Medicine*, 47: 601–606.

[20] Tofts, P. S., Brix, G., Buckley, D. L., Evelhoch, J. L., Henderson, E., Knopp, M. V., Larsson, H. B. W., Lee, T. Y., Mayr, N. A., Parker, G. J. M., Port, R. E., Taylor, J. and Weisskoff, R. M. (1999). Estimating kinetic parameters from dynamic contrast-enhanced T_1-weighted MRI of a diffusible tracer: standardized quantities and symbols. *Journal of Magnetic Resonance Imaging*, 10: 223–232.

[21] Renkin, E. M. (1952). Capillary permeability to lipid-soluble molecule. *American Journal of Physiology*, 168: 538–545.

[22] Crone, C. (1963). The permeability of capillaries in various organs as determined by use of the "indicator diffusion" method. *Acta Physiologica Scandinavica*, 58: 292–305.

[23] Murase, K. (2004). Efficient method for calculating kinetic parameters using T1-weighted dynamic contrast-enhanced magnetic resonance imaging. *Magnetic Resonance in Medicine*, 51: 858–862.

[24] Huang, S. C., Phelps, M. E., Hoffman, E. J., Sideris, K., Selin, C. J. and Kuhl, D. E. (1980). Noninvasive determination of local cerebral metabolic rate of glucose in man. *American Journal of Physiology*, 238: E69–E82.

[25] Murase, K. and Tanki, N. (2011). Numerical solutions to the time-dependent Bloch equations revisited. *Magnetic Resonance Imaging*, 29: 126–131.

[26] Miles, K. A., Hayball, M. P. and Dixon, A. K. (1993). Functional images of hepatic perfusion obtained with dynamic CT. *Radiology*, 188: 405–411.

[27] Materne, R., Van Beers, B. E., Smith, A. M., Leconte, I., Jamart, J., Dehoux, J. P., Keyeux, A. and Horsmans, Y. (2000). Non-invasive quantification of liver perfusion with dynamic computed tomography and a dual input one-compartmental model. *Clinical Science*, 99: 517–525.

[28] Leggett, D. A., Kelley, B. B., Bunce, I. H. and Miles, K. A. (1997). Colorectal cancer: diagnostic potential of CT measurements of

hepatic perfusion and implications for contrast enhancement protocols. *Radiology*, 205: 716–720.

[29] Becker, G. A., Muller-Schauenburg, W., Spilker, M. E., Machulla, H. J. and Piert, M. (2005). A priori identifiability of a one-compartment model with two input functions for liver blood flow measurements. *Physics in Medicine and Biology*, 50: 1393–1404.

[30] Pandharipande, P. V., Krinsky, G. A., Rusinek, H. and Lee, V. S. (2005). Perfusion imaging of the liver: current challenges and future goals. *Radiology*, 234: 661–673.

[31] Munk, O. L., Bass, L., Roelsgaard, K., Bender, D., Hansen, S. B. and Keiding, S. (2001). Liver kinetics of glucose analogs measured in pigs by PET: importance of dual-input blood sampling. *Journal of Nuclear Medicine*, 42: 795–801.

[32] Miyazaki, S., Murase, K., Yoshikawa, T., Morimoto, S., Ohno, Y. and Sugimura, K. (2008). A quantitative method for estimating hepatic blood flow using a dual-input single-compartment model. *British Journal of Radiology*, 81: 790–800.

[33] Murase, K., Miyazaki, S. and Yang, X. (2007). An efficient method for calculating kinetic parameters in a dual-input single-compartment model. *British Journal of Radiology*, 80: 371–375.

[34] Eastwood, J. D., Lev, M. H., Azhari, T., Lee, T. Y., Barboriak, D. P., Delong, D. M., Fitzek, C., Herzau, M., Wintermark, M., Meuli, R., Brazier, D. and Provenzale, J. M. (2002). CT perfusion scanning with deconvolution analysis: pilot study in patients with acute middle cerebral artery stroke. *Radiology*, 222: 227–236.

[35] George, R. T., Jerosch-Herold, M., Silva, C., Kitagawa, K., Bluemke, D. A., Lima, J. A. C. and Lardo, A. C. (2007). Quantification of myocardial perfusion using dynamic 64-detector computed tomography. *Investigative Radiology*, 42: 815–822.

[36] Won, C., Chon, D., Tajik, J., Tran, B. Q., Robinswood, G. B., Beck, K. C. and Hoffman, E. A. (2003). CT-based assessment of regional pulmonary microvascular blood flow parameters. *Journal of Applied Physiology*, 94: 2483–2493.

[37] Miles, K. A., Lee, T. Y., Goh, V., Klotz, E., Cuenod, C., Bisdas, S., Groves, A. M., Hayball, M. P., Alonzi, R. and Brunner, T. (2012). Current status and guidelines for the assessment of tumor vascular support with dynamic contrast-enhanced computed tomography. *European Radiology*, 22: 1430–1441.

[38] Kusakabe, Y., Miyazaki, S., Tachibana, A., Matsuura, R., Matsuura, N. and Murase, K. (2009). Development of a method to quantitatively monitor the effect of inhibition of nitric oxide synthase on tumor vascular activity using dynamic contrast-enhanced computed tomography. *Journal of Medical Engineering and Technology*, 33: 460–469.

[39] Murase, K., Shinohara, M. and Yamazaki, Y. (2001). Accuracy of deconvolution analysis based on singular value decomposition for quantification of cerebral blood flow using dynamic susceptibility contrast-enhanced magnetic resonance imaging. *Physics in Medicine and Biology*, 46: 3147–3159.

[40] Murase, K., Yamazaki, Y. and Miyazaki, S. (2004). Deconvolution analysis of dynamic contrast-enhanced data based on singular value decomposition optimized by generalized cross validation. *Magnetic Resonance in Medical Sciences*, 3: 165–175.

[41] Hansen, P. C. (1992). Analysis of discrete ill-posed problems by means of the L-curve. *SIAM Review*, 34: 561–580.

[42] Tobler, H.J. and Engel, G. (1983). Affinity spectra: a novel way for the evaluation of equilibrium binding experiments. *Naunyn Schmiedebergs Arch Pharmacol,* 322: 183–192.

[43] Cunningham, V. J. and Jones, T. (1993). Spectral analysis of dynamic PET studies. *Journal of Cerebral Blood Flow and Metabolism*, 13: 15–23.

[44] Lawson, C. L. and Hanson, R. J. (1974). *Solving Least Squares Problems*, Prentice-Hall, Englewood Cliffs, New Jersey.

[45] Murase, K. (2003). Spectral analysis: principle and clinical applications. *Annals of Nuclear Medicine*, 17: 427–434.

[46] Fukui, A., Murase, K., Tsuda, T., Fujii, T. and Ikezoe, J. (2000). Assessment of liver function in chronic liver diseases and regional

function of irradiated liver by means of ^{99m}Tc-galactosyl human serum albumin liver scintigraphy and quantitative spectral analysis. *Annals of Nuclear Medicine*, 14: 467–476.

[47] Murase, K., Yamazaki, Y., Mochizuki, T. and Ikezoe, J. (2002). Renal uptake rate measurement of ^{99m}Tc-dimercaptosuccinic acid using spectral analysis. *Nuclear Medicine Communications*, 23: 265–273.

[48] Murase, K., Inoue, T., Fujioka, H., Ishimaru, Y., Akamune, K., Yamamoto, Y., Mochizuki, T. and Ikezoe, J. (1999). An alternative approach to estimation of the brain perfusion index for measurement of cerebral blood flow using technetium-99m compounds. *European Journal of Nuclear Medicine*, 26: 1333–1339.

[49] Ahearn, T. S., Staff, R. T., Redpath, T. W. and Semple, S. I. K. (2005). The use of the Levenberg-Marquardt curve-fitting algorithm in pharmacokinetic modelling of DCE-MRI data. *Physics in Medicine and Biology*, 50: N85–N92.

[50] Nelder, J. A. and Mead, R. (1965). A simplex method for function minimization. *Computer Journal*, 7: 308–313.

[51] Press, W. H., Teukolsky, S. A., Vetterling, W. T. and Flannery, B. P. (1992). *Numerical Recipes in C*, Cambridge University Press, New York.

[52] Glatting, G., Kletting, P., Reske, S. N., Hohl, K. and Ring, C. (2007). Choosing the optimal fit function: comparison of the Akaike information criterion and the F-test. *Medical Physics*, 34: 4285–4292.

[53] Blomqvist, G. (1984). On the construction of functional maps in positron emission tomography. *Journal of Cerebral Blood Flow and Metabolism*, 4: 629–632.

[54] Gjedde, A. (1981). High- and low-affinity transport of D-glucose from blood to brain. *Journal of Neurochemistry*, 36: 1463–1471.

[55] Patlak, C. S. and Blasberg, R. G. (1985). Graphical evaluation of blood-to-brain transfer constants from multiple-time uptake data. Generalizations. *Journal of Cerebral Blood Flow and Metabolism*, 5: 584–590.

[56] Fan, X., Medved, M., River, J. N., Zamora, M., Corot, C. and Robert, P. (2004). New model for analysis of dynamic contrast-enhanced MRI

data distinguishes metastatic from nonmetastatic transplanted rodent prostate tumors. *Magnetic Resonance in Medicine*, 51: 487–494.

[57] Fan, X., Medved, M., Karcamar, G. S., Yang, C., Foxley, S., Arkani, S., Recant, W., Zamora, M. A., Abe, H. and Newstead, G. M. (2007). Diagnosis of suspicious breast lesions using an empirical mathematical model for dynamic contrast-enhanced MRI. *Magnetic Resonance Imaging*, 25: 593–603.

[58] Oto, A., Fan, X., Mustafi, D., Jansen, S. A., Karczmar, G. S., Rubin, D. T. and Kayhan, A. (2009). Quantitative analysis of dynamic contrast enhanced MRI for assessment of bowel inflammation in Crohn's disease. *Academic Radiology*, 16: 1223–1230.

[59] Ba-Ssalamah, A., Uffmann, M., Saini, S., Bastani, N., Herold, C. and Schima, W. (2009). Clinical value of MRI liver-specific contrast agents: a tailored examination for a confident non-invasive diagnosis of focal liver lesions. *European Radiology*, 19: 342–357.

[60] Reimer, P., Rummeny, E. J. and Daldrup, H. E. (1997). Enhancement characteristics of liver metastases, hepatocellular carcinoma, and hemangiomas with Gd-EOB-DTPA: preliminary results with dynamic MR imaging. *European Radiology*, 7: 275–280.

[61] Tsuda, N., Okada, M. and Murakami, T. (2007). Potential of gadolinium-ethoxybenzyl-diethylenetriamine pentaacetic acid (Gd-EOB-DTPA) for differential diagnosis of nonalcoholic steatohepatitis and fatty liver in rats using magnetic resonance imaging. *Investigative Radiology*, 42: 242–247.

[62] Moriyama, Y., Kobayashi, S., Ogihara, R., Koto, D., Kitamura, A., Matsushita, T., Nishiura, M. and Murase, K. (2012). Evaluation of concanavalin A-induced acute liver injury in rats using an empirical mathematical model and dynamic contrast-enhanced MRI with Gd-EOB-DTPA. *Magnetic Resonance in Medical Sciences*, 11: 53–60.

[63] Hayton, P., Brady, M., Tarassenko, L. and Moore, N. (1997). Analysis of dynamic MR breast images using a model of contrast enhancement. *Medical Image Analysis*, 1: 207–224.

In: Mathematical Modeling ISBN: 978-1-53613-162-8
Editors: Seth Sparks and Bill Willis

Chapter 3

How Are Groceries, Images, and Matrices Related?

Joan Gómez Urgellés*, PhD
Department of Mathematics
Polytechnic University of Catalonia EPSEVG
Vilanova i la Geltrú. Spain

Abstract

My goal is to establish a relation between Mathematics and reality.

Actually, from daily matters, I am demonstrating that these situations can be interpreted building Mathematical Models which bring information about all of them, in this case Algebra of Matrices.

Firstly, a matrix is introduced as a Mathematical Model of an image and then, the subtraction or the difference of matrices shows a new way to compare images. To reach this model is necessary a specific software (Octave), consequently the digital competence become part of this work.

---

* Corresponding Author Email: joan.vicenc.gomez@upc.edu.

Secondly, naturally the concept of the Scaled Euclidean Product has been developed, also the product of matrices as Mathematical Models of Matrix Algebra for daily situations related to Supermarket shopping. Consequently, these daily situations lead to introduce Matrix Models linking quantities, prices and bought products; a concept which allows a transverse and interdisciplinary branch of the Mathematics.

With the proposed examples is possible to assure that the Mathematical Modeling is a valid tool for teaching / learning in a competent way, at the same time recommendable to be included in the academic curricula, reaching a transverse view of the Mathematics Actually, it materializes epistemological, heuristic and cognitive aspects in the context of the Mathematics teaching.

Synthesizing we can consider mathematical modeling as an efficient teaching innovation tool, and a transmission strap that provides the knowledge approaching mathematics to the reality.

1. Introduction

This article is a proposition for the teaching / learning of some matrix calculation elements from mathematical modeling. As a matter of fact, some daily situations are established by having their matrices and their operations as a mathematical model as well, in particular by showing how we can create models to illustrate the matrix concept and also by introducing basic operations of difference and product of matrices.

Firstly, a matrix is shown as a mathematical model of an image and then how the matrix difference becomes a model for image comparison is discussed.

However, to do this task software such as Octave (or similar software) is necessary. This tool allows the research of a numerical model of a black and white image represented by a matrix.

Furthermore, we see how the matrix product is a model which can be naturally deduced from the grocery shopping routine. The main idea is to underline the matrix calculation epistemology in order to reinforce the students' cognitive character, bringing a contextual view of daily matters in real life at the same time, enriching the heuristic, thus allowing the visualization of the connection among the mathematical symbolism (introduced on the model) and the real situations.

1.1. Working with Images: The Matrices Difference as a Mathematical Model

1.1.1. A Matrix as a Mathematical Model of a Black and White Image

When we talk about images, mathematics has an important role. Actually, technically each image can be seen as a table of numbers (formally known as a matrix). Then, defining "an image composed by composition of M per N pixels", it means that it can be represented by a matrix with M rows and N columns, generally with values between 0 and 255 (256 elements). The number of pixels is called "resolution." The procedure to obtain the matrix has been done through very sophisticated mathematical algorithms implemented by the software (or MatLab at a cost of 50 $ for students or 150 $ for home users).

When we say 15 per 13 pixels, we mean something similar to the figure below:

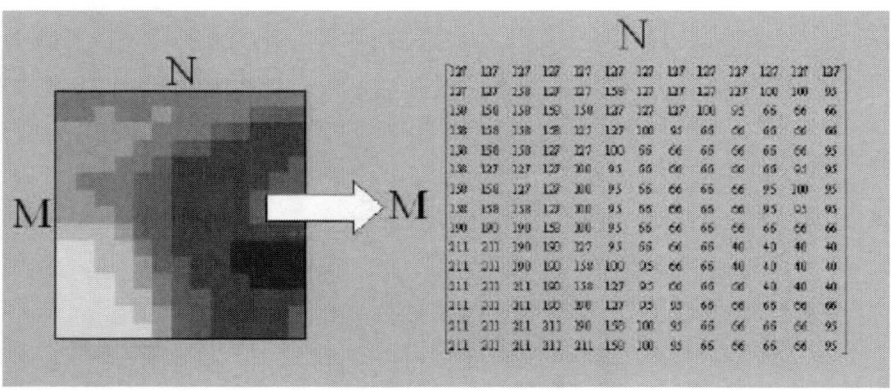

Figure 1. The pixels and their correspondent array.

When we talk about "5 megapixels," we are really talking about 5 million pixels. However, if we read "640 x 480," that means 640 columns per 480 rows matrix.

Now we are analyzing a real situation. Consider these violin pictures:

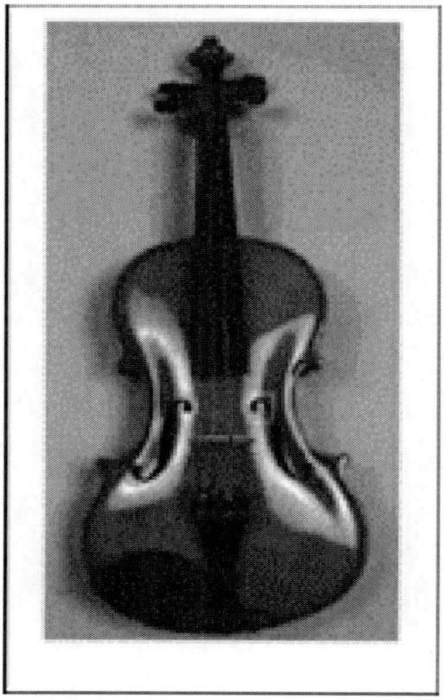

Figure 2. Violin.

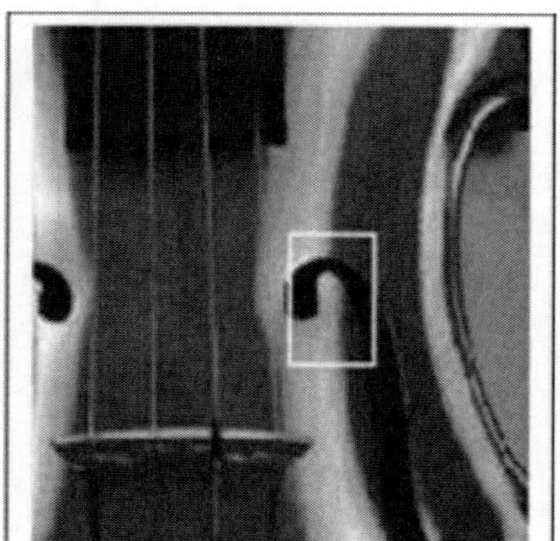

Figure 3. Violin detail.

This violin picture image matches the Matrix below:

Table 1. Matrix of the Violin

170	170	180	181	186	189	191	186	179	172	164	152	98	56	57	53	50	55	50	44	45	41	41	45	46	47
168	175	177	177	188	194	187	186	179	169	161	153	93	59	55	49	50	49	47	43	44	46	44	45	43	40
169	171	182	183	190	194	192	187	172	171	158	136	90	62	53	55	52	48	49	42	45	48	43	40	43	41
175	175	185	192	195	194	193	187	174	172	164	150	87	55	52	51	48	47	48	47	46	42	40	43	43	43
176	181	185	189	191	192	195	190	178	174	165	145	85	54	48	47	46	48	49	49	42	45	41	42	43	43
178	180	189	194	197	197	196	189	181	174	164	146	73	53	58	48	46	45	45	50	46	46	43	41	43	43
182	186	192	197	200	201	198	190	184	184	176	156	74	46	48	45	45	39	41	42	43	44	42	40	41	43
182	189	195	200	209	217	215	205	190	165	142	116	79	49	45	43	43	42	46	47	43	42	41	41	39	44
183	187	200	211	236	223	140	56	28	16	11	7	12	13	23	38	49	48	46	46	45	41	40	38	40	43
191	191	220	214	143	43	9	5	7	7	8	6	8	7	7	12	23	33	44	43	44	42	42	40	42	43
196	227	190	45	5	6	7	9	8	7	6	7	4	9	7	6	7	11	24	47	37	39	41	44	43	41
215	182	25	4	8	8	9	7	9	6	7	46	44	7	8	8	6	8	9	19	31	38	40	42	45	41
180	21	6	8	9	8	9	9	2	25	122	178	164	39	9	7	9	7	7	8	15	30	38	38	41	38
63	2	8	8	7	9	8	4	22	142	203	179	163	126	50	8	6	8	7	9	6	16	33	38	39	41
11	8	9	8	8	8	6	31	141	183	187	181	168	167	112	35	9	8	8	7	9	8	14	31	40	40
5	9	8	8	7	8	9	117	172	169	187	182	163	158	131	65	21	7	7	7	6	8	8	18	37	37
8	9	8	8	9	7	6	62	174	172	183	180	165	149	127	85	33	12	8	7	8	8	8	9	21	37
10	7	8	8	8	7	7	12	134	181	181	174	160	153	128	88	41	26	8	8	8	7	7	7	8	27
9	8	8	8	7	7	7	7	68	182	176	170	164	154	119	93	41	35	17	8	7	8	7	7	7	14
9	9	7	8	8	7	7	7	21	152	176	168	160	151	126	87	41	39	29	8	6	7	7	8	7	8
9	8	9	8	8	7	8	7	10	117	182	166	150	147	120	75	39	39	35	16	6	8	7	6	7	6
10	9	9	7	8	8	7	8	8	108	177	167	159	151	105	82	42	39	37	27	8	8	7	7	7	8
10	9	8	8	7	8	8	8	11	122	172	160	152	152	117	80	49	42	37	33	13	6	8	7	8	7
10	10	9	10	8	8	8	6	26	146	168	160	150	144	110	82	44	38	39	34	18	8	7	8	7	8
7	10	10	9	9	9	8	6	97	167	157	161	149	147	102	66	49	42	36	34	25	7	8	6	8	8
35	9	7	9	8	7	4	42	151	161	162	160	153	150	105	77	43	40	35	35	30	9	7	8	7	6
115	50	24	12	13	26	66	142	157	155	163	160	152	148	107	67	48	40	35	34	31	12	9	7	7	8
140	123	117	95	92	129	154	153	158	159	162	157	151	148	105	62	44	38	38	34	32	17	7	8	6	8
136	134	145	146	143	152	151	157	166	167	166	160	153	147	112	78	43	39	41	36	33	23	7	7	9	7
138	142	144	144	150	153	156	158	168	171	171	157	149	155	108	61	51	43	40	37	34	24	8	7	6	9
143	141	145	149	153	159	159	157	167	176	170	163	160	154	110	62	48	43	38	35	32	30	12	8	8	7
148	150	145	151	156	159	158	162	169	172	167	162	156	153	114	59	53	45	38	35	30	33	17	7	7	8
150	150	150	157	164	164	166	170	174	172	174	166	155	151	116	71	51	47	41	36	34	33	21	8	9	9
145	148	153	158	160	168	168	171	178	178	170	165	161	153	104	66	47	43	40	37	34	33	21	8	8	8
151	151	156	157	163	166	165	174	176	177	170	165	163	154	118	67	45	41	40	35	33	34	25	9	7	8
147	152	156	161	167	169	167	171	177	180	180	167	154	155	128	80	52	38	37	37	34	34	26	10	7	8
154	153	158	164	170	169	171	172	185	181	173	171	159	156	128	81	60	43	42	40	36	33	27	10	8	8
150	155	162	167	168	172	172	176	179	180	175	169	163	155	133	77	54	44	44	38	38	33	29	11	8	8

Take a glance at this unbelievable Numeric Table, even the density and the placement of the numbers "drawing" the violin profile. Next, the procedure in order to get the model is fully explained.

Actually, we can choose any image in our computer, by selecting with the cursor over the image, right clicking it, and then opening a Properties Window that shows all of the information about the image size. Depending on the resolution and the available space on the disk, it is possible to save the image in different formats such as BMP, TIFF, or JPEG as well.

In order to get a matrix model of an image, one needs software such as Octave. Formally Octave works with sophisticated numeric methods for obtaining the Matrix Model. The procedures are explained below.

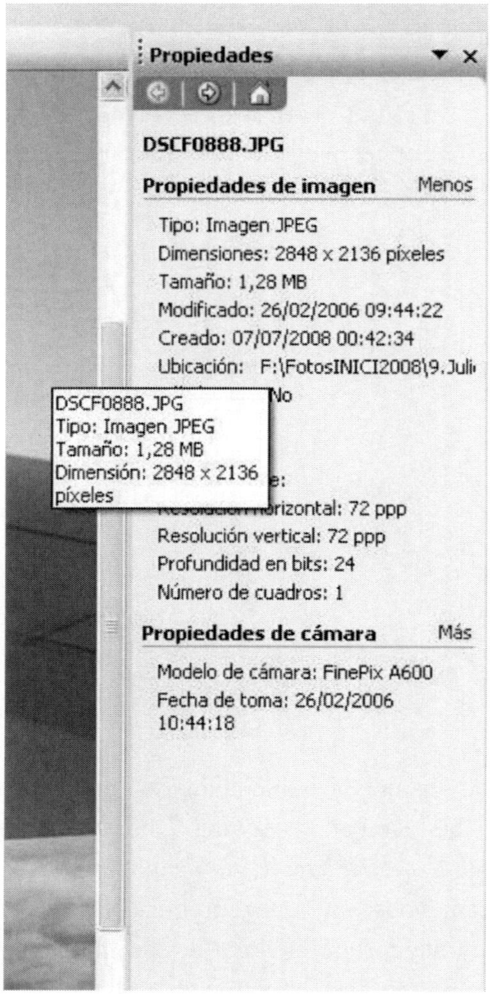

Figure 4. Octave Properties Window.

1.1.2. Octave: Generating the Pixels Matrix of an Image

Octave is a free program available for Windows, Mac, and Linux, developed for Numeric Calculations. It is available on the web https://www.gnu.org/software/octave/ it was developed around 1988, created by Chemical Engineering students from Texas University and Wisconsin-Madison University to be applied to support Chemical Reactors drawing. Actually, Octave is a free option to the well-known MathLab.

Octave has a wide kit of tools to solve algebra, calculation, and statistics problems. It is also able to process digital images. A Smartphone version is available.

We are working with Black and White images because the associated Matrix is bi-dimensional, which means that it is a Numbering Table with rows and columns. On the other hand, in case of colored images, a "three-dimensional Matrix" would be generated, and each color would be obtained from the basic RGB (Red, Green, Blue).

With Octave, we can generate the Pixel Matrix of any image. How can we do that? By following the steps below:

1.1.2.1. To Install the Program, Go to the Link:

https://www.gnu.org/software/octave/ or
ftp://ftp.gnu.org/gnu/octave/windows/octave-4.0.0_0-installer.exe

1.1.2.2. Once the Octave Installation Is Completed, Run the Program by Opening a Similar Window Such As

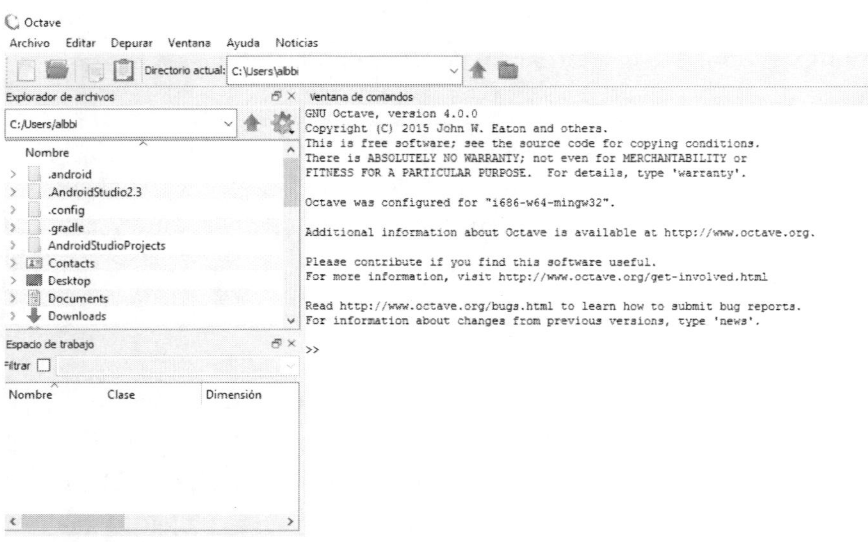

Figure 5. Running Octave.

1.1.2.3. Choose a Previously Saved Image in the Directory

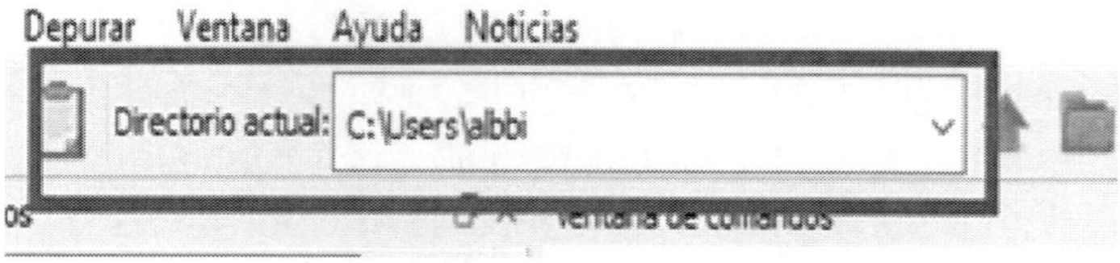

Figure 6. Saving in the directory.

1.1.2.4. And Then, Select the Image Whose Pixel Matrix Associated We Want to Know

Remember that it is necessary to save the image in the folder created by Octave, which in our case is *C:\Users\albbi.*

Literally, in the Octave window we should go to the directory in which the images have been saved.

With the image selected in the folder, introduce the following instruction in the command line:

image = imread ('image name. extension')

Summarizing: We can introduce the concept of Matrix as a model of a Black and White image.

2. Modeling Experience

2.1. Introduction

Now, we are doing an experience which has been explained in the classroom by showing the Matrix difference as a Mathematical Model of an image. In our example, we are considering two different black and which images, previously saved in our computer, and comparing both of them.

In our computer, these images have been saved as "image1.jpg," "image2.jpg."

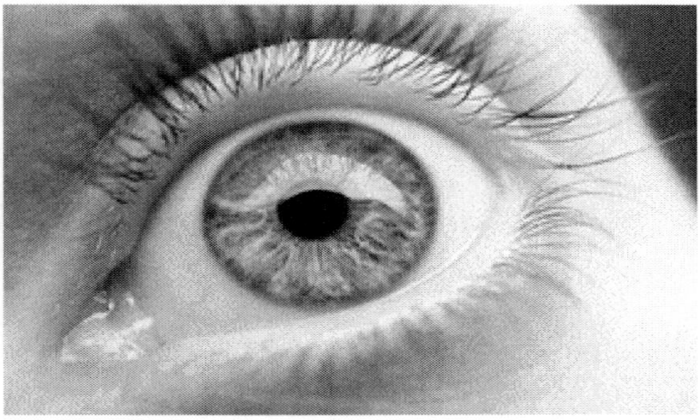

Figure 7. Image 1.

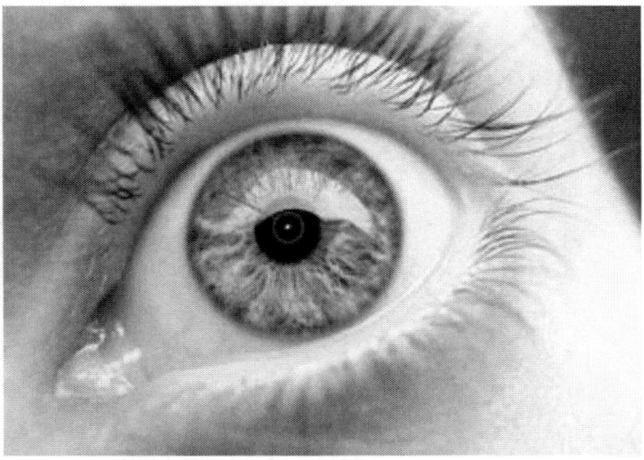

Figure 8. Image 2.

Apparently, these pictures seem identical, but this is not true; the second picture has a small different dot in yellow in order to be easily seen.

Now we are introducing the following lines in octave command:

Variable Name I = imread (“fotografia1.jpg”);

```
>> I = imread("fotografia1.jpg");
```

Figure 9. An Octave command line for the variable I.

As shown above, “I” has been chosen as a Variable Name.

In this line, we are saving the Variable “I,” the “image 1.” However, we are really saving the Matrix of picture 1.

Next, we are introducing the command Variable Name 2 = I2 = imread ("fotografia2.jpg");

```
>> I2 = imread("fotografia2.jpg");
```

Figure 10. An Octave command line for the variable I2.

Saving the variable "I2" ("picture 2")

By pressing "Enter" after each instruction, the window shows respectively the Matrix as a Mathematical Model of each image respectively obtaining:

Table 2. 1st Image Partial view (image 1)

```
>> I
I =

 Columns 1 through 21:

  159  159  159  159  159  159  159  159  161  16
  161  161  161  161  161  161  161  161  159  15
  161  161  161  161  161  161  161  161  157  15
  159  159  159  159  159  159  159  159  157  15
  158  158  158  158  158  158  158  158  158  15
  158  158  158  158  158  158  158  158  158  15
  158  158  158  158  158  158  158  158  156  15
  157  157  157  157  157  157  157  157  154  15
  155  155  155  155  155  155  155  155  157  15
  155  155  155  155  155  155  155  155  155  15
  156  156  156  156  156  156  156  156  154  15
  155  155  155  155  155  155  155  155  153  15
  154  154  154  154  154  154  154  154  153  15
```

Table 3. Full Capture of the "image 1" Matrix

159	159	159	159	159	159	159	159	161	161	162	163	164	164	165	165	166	166	167	168	169
161	161	161	161	161	161	161	161	159	159	160	161	162	162	163	163	164	165	165	166	167
161	161	161	161	161	161	161	161	157	158	158	159	160	160	161	161	163	164	164	165	166
159	159	159	159	159	159	159	159	157	157	158	159	160	160	161	161	164	164	165	166	167
158	158	158	158	158	158	158	158	158	158	159	159	160	161	162	162	165	165	166	167	168
158	158	158	158	158	158	158	158	158	158	159	159	160	161	161	162	164	164	165	166	167
158	158	158	158	158	158	158	158	156	156	157	157	158	159	160	160	161	162	162	163	165
157	157	157	157	157	157	157	157	154	154	155	155	156	157	158	158	159	159	160	161	162
155	155	155	155	155	155	155	155	157	157	158	158	159	159	159	160	164	164	164	164	164
155	155	155	155	155	155	155	155	155	156	156	156	157	157	158	158	162	162	162	162	162
156	156	156	156	156	156	156	156	154	154	154	155	155	156	156	157	159	159	159	159	159
155	155	155	155	155	155	155	155	153	153	154	154	155	156	156	156	156	156	157	157	157
154	154	154	154	154	154	154	154	153	153	154	155	155	156	156	157	155	155	155	156	156
152	152	152	152	152	152	152	152	151	152	152	153	154	155	155	155	153	154	154	154	155
150	150	150	150	150	150	150	150	148	148	149	150	151	151	152	152	151	151	152	152	152
148	148	148	148	148	148	148	148	145	146	146	147	148	149	149	150	149	150	150	150	151
147	147	147	148	148	149	149	149	148	147	147	147	147	147	147	148	148	148	148	149	149
146	147	147	147	148	148	149	149	147	146	146	145	145	146	146	147	148	148	148	149	149
146	146	146	147	147	147	147	148	146	145	145	144	144	145	145	146	147	147	148	148	149
146	146	146	146	146	146	147	147	146	145	145	145	145	145	145	146	146	146	147	147	148
146	146	146	146	146	146	146	146	147	146	146	145	145	146	146	147	145	145	146	146	147
147	147	147	147	147	147	147	147	147	147	146	146	146	146	147	147	145	145	146	146	147
148	148	148	148	147	147	147	147	146	146	145	145	145	145	146	146	146	146	147	147	148
149	149	148	148	148	148	148	147	145	144	144	143	143	144	144	145	147	147	148	148	149
149	148	148	147	146	145	145	145	143	143	143	143	143	143	143	143	146	146	146	147	147
149	149	148	147	146	146	145	145	143	143	143	143	143	143	143	143	145	146	146	147	147
149	149	148	147	146	146	145	145	143	143	143	143	143	143	143	143	145	145	146	146	147
148	148	148	147	146	145	145	144	143	143	143	143	143	143	143	142	145	145	145	146	146

Analogically for the second image:

Table 4. 2nd Image Partial view (image 2)

```
>> I2
I2 =

ans(:,:,1) =

 Columns 1 through 18:

  159  159  159  159  159  159  159  159  161  161  1
  161  161  161  161  161  161  161  161  159  159  1
  161  161  161  161  161  161  161  161  157  157  1
  159  159  159  159  159  159  159  159  157  157  1
  158  158  158  158  158  158  158  158  158  158  1
  158  158  158  158  158  158  158  158  158  158  1
  158  158  158  158  158  158  158  158  156  156  1
  157  157  157  157  157  157  157  157  154  154  1
  155  155  155  155  155  155  155  155  157  157  1
  155  155  155  155  155  155  155  155  155  156  1
  156  156  156  156  156  156  156  156  154  154  1
  155  155  155  155  155  155  155  155  153  153  1
  154  154  154  154  154  154  154  154  153  153  1
  152  152  152  152  152  152  152  152  151  152  1
  150  150  150  150  150  150  150  150  148  148  1
```

Table 5. Full Capture of the 2nd Image Matrix

159	159	159	159	159	159	159	159	161	161	162	162	163	164	165	165	166	166	167	168	169
161	161	161	161	161	161	161	161	159	159	160	161	161	162	163	163	164	165	165	166	167
161	161	161	161	161	161	161	161	157	157	158	159	160	160	161	161	163	164	164	165	166
159	159	159	159	159	159	159	159	157	157	158	159	159	160	161	161	164	164	165	166	167
158	158	158	158	158	158	158	158	158	158	159	160	160	161	162	162	165	165	166	167	168
158	158	158	158	158	158	158	158	158	158	159	159	160	161	162	162	164	164	165	166	167
158	158	158	158	158	158	158	158	156	156	157	158	158	159	160	160	161	162	162	163	165
157	157	157	157	157	157	157	157	154	154	155	156	157	157	158	158	159	159	160	161	162
155	155	155	155	155	155	155	155	157	157	158	158	159	159	159	160	164	164	164	164	164
155	155	155	155	155	155	155	155	155	156	156	156	157	157	158	158	162	162	162	162	162
156	156	156	156	156	156	156	156	154	154	154	155	155	156	156	157	159	159	159	159	159
155	155	155	155	155	155	155	155	153	153	154	154	155	156	156	156	156	157	157	157	157
154	154	154	154	154	154	154	154	153	153	154	155	155	156	156	157	155	155	155	156	156
152	152	152	152	152	152	152	152	151	152	152	153	154	155	155	155	153	153	154	154	155
150	150	150	150	150	150	150	150	148	148	149	150	151	151	152	152	151	151	151	152	153
148	148	148	148	148	148	148	148	145	146	146	147	148	149	149	150	149	149	150	150	151
147	147	147	148	148	149	149	149	148	147	147	147	147	147	147	148	148	148	148	149	149
146	147	147	147	148	148	149	149	147	146	146	145	145	146	146	147	148	148	148	149	149
146	146	146	147	147	147	147	148	146	145	145	144	144	145	145	146	147	147	148	148	149
146	146	146	146	146	146	147	147	146	145	145	145	145	145	145	146	146	146	147	147	148
146	146	146	146	146	146	146	146	147	146	146	145	145	146	146	147	145	145	146	146	147
147	147	147	147	147	147	147	147	147	147	146	146	146	146	147	147	145	145	146	146	147
148	148	148	148	147	147	147	147	146	146	145	145	145	145	146	146	146	146	147	147	148
149	149	148	148	148	148	148	147	145	144	144	143	143	144	144	145	147	147	148	148	149
149	148	148	147	146	145	145	145	143	143	143	143	143	143	143	143	146	146	146	147	147
149	149	148	147	146	146	145	145	143	143	143	143	143	143	143	143	145	146	146	147	147
149	149	148	147	146	146	145	145	143	143	143	143	143	143	143	143	145	145	146	146	147
148	148	148	147	146	145	145	144	143	143	143	143	143	143	143	142	145	145	145	146	146

We realize that the difference between the images reflects in the different Matrix Model values.

Strictly speaking: we can get the differences between images by subtracting these matrices and concluding that the regions with zeros do not have changes. On the other hand, the regions with values different from zero mean that they do have changes.

Summarizing: The difference between Matrices would be a Mathematical Modeling to compare images.

The Model has been applied in the class to the First Course of Computer Science at EPSEVG University, as a group work developed by students, despite the fact that they had never worked with Matrices before. However, they were able to explain the work in the classroom to their other classmates, as shown in the picture below.

Figure 11. Computer Science University class.

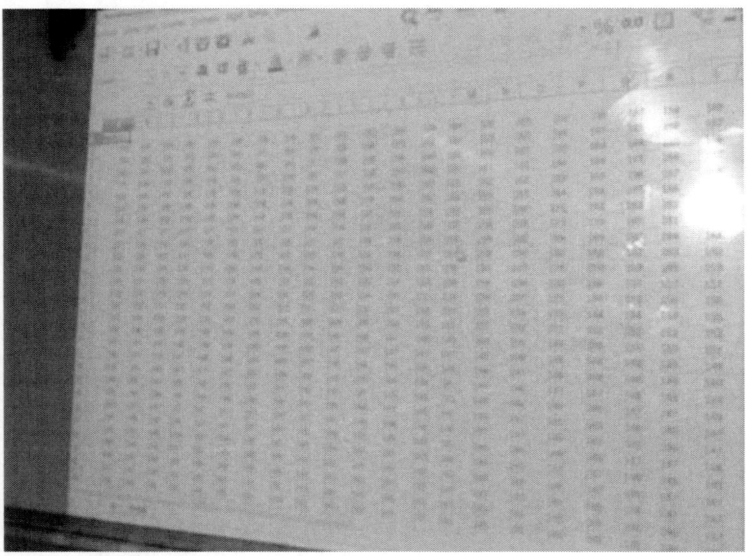

Figure 12. The Matrix obtained by the Computer Science students.

Octave also allows to subtract the obtained Matrices. It's even possible to select the Matrices by attaching them to the Excel database and then subtracting them. In our experience:

Table 6. Matrix 1

1	Matriu1																				
2	159	159	159	159	159	159	159	159	161	161	162	163	164	164	165	165	166	166	167	168	169
3	161	161	161	161	161	161	161	161	159	159	160	161	162	162	163	163	164	165	165	166	167
4	161	161	161	161	161	161	161	161	157	158	158	159	160	160	161	161	163	164	164	165	166
5	159	159	159	159	159	159	159	159	157	157	158	159	160	160	161	161	164	164	165	166	167
6	158	158	158	158	158	158	158	158	158	158	159	159	160	161	162	162	165	165	166	167	168
7	158	158	158	158	158	158	158	158	158	158	159	159	160	161	161	162	164	164	165	166	167
8	158	158	158	158	158	158	158	158	156	156	157	157	158	159	160	160	161	162	162	163	165
9	157	157	157	157	157	157	157	157	154	154	155	155	156	157	158	158	159	159	160	161	162
10	155	155	155	155	155	155	155	155	157	157	158	158	159	159	159	160	164	164	164	164	164
11	155	155	155	155	155	155	155	155	155	156	156	156	157	157	158	158	162	162	162	162	162
12	156	156	156	156	156	156	156	156	154	154	154	155	155	156	156	157	159	159	159	159	159
13	155	155	155	155	155	155	155	155	153	153	154	154	155	156	156	156	156	156	157	157	157
14	154	154	154	154	154	154	154	154	153	153	154	155	155	156	156	157	155	155	155	156	156
15	152	152	152	152	152	152	152	152	151	152	152	153	154	155	155	155	153	154	154	154	155
16	150	150	150	150	150	150	150	150	148	148	149	150	151	151	152	152	151	151	152	152	152
17	148	148	148	148	148	148	148	148	145	146	146	147	148	149	149	150	149	150	150	150	151
18	147	147	147	148	148	149	149	149	148	147	147	147	147	147	147	148	148	148	148	149	149
19	146	147	147	147	148	148	149	149	147	146	146	145	145	146	146	147	148	148	148	149	149
20	146	146	146	147	147	147	147	148	146	145	145	144	144	145	145	146	147	147	148	148	149
21	146	146	146	146	146	146	147	147	146	145	145	145	145	145	145	146	146	146	147	147	148
22	146	146	146	146	146	146	146	146	147	146	146	145	145	146	146	147	145	145	146	146	147
23	147	147	147	147	147	147	147	147	147	147	146	146	146	146	147	147	145	145	146	146	147
24	148	148	148	148	147	147	147	147	146	146	145	145	145	145	146	146	146	146	147	147	148
25	149	149	148	148	148	148	148	147	145	144	144	143	143	144	144	145	147	147	148	148	149
26	149	148	148	147	146	145	145	145	143	143	143	143	143	143	143	143	146	146	146	147	147
27	149	149	148	147	146	146	145	145	143	143	143	143	143	143	143	143	145	146	146	147	147
28	149	149	148	147	146	146	145	145	143	143	143	143	143	143	143	143	145	145	146	146	147
29	148	148	148	147	146	145	145	144	143	143	143	143	143	143	143	142	145	145	145	146	146

Table 7. Matrix 2

32	Matriu2																				
33	159	159	159	159	159	159	159	159	161	161	162	162	163	164	165	165	166	166	167	168	169
34	161	161	161	161	161	161	161	161	159	159	160	161	161	162	163	163	164	165	165	166	167
35	161	161	161	161	161	161	161	161	157	157	158	159	160	160	161	161	163	164	164	165	166
36	159	159	159	159	159	159	159	159	157	157	158	159	159	160	161	161	164	164	165	166	167
37	158	158	158	158	158	158	158	158	158	158	159	160	160	161	162	162	165	165	166	167	168
38	158	158	158	158	158	158	158	158	158	158	159	159	160	161	162	162	164	164	165	166	167
39	158	158	158	158	158	158	158	158	156	156	157	158	158	159	160	160	161	162	162	163	165
40	157	157	157	157	157	157	157	157	154	154	155	156	157	157	158	158	159	159	160	161	162
41	155	155	155	155	155	155	155	155	157	157	158	158	159	159	159	160	164	164	164	164	164
42	155	155	155	155	155	155	155	155	155	156	156	156	157	157	158	158	162	162	162	162	162
43	156	156	156	156	156	156	156	156	154	154	154	155	155	156	156	157	159	159	159	159	159
44	155	155	155	155	155	155	155	155	153	153	154	154	155	156	156	156	156	157	157	157	157
45	154	154	154	154	154	154	154	154	153	153	154	155	155	156	156	157	155	155	155	156	156
46	152	152	152	152	152	152	152	152	151	152	152	153	154	155	155	155	153	153	154	154	155
47	150	150	150	150	150	150	150	150	148	148	149	150	151	151	152	152	151	151	151	152	153
48	148	148	148	148	148	148	148	148	145	146	146	147	148	149	149	150	149	149	150	150	151
49	147	147	147	148	148	149	149	149	148	147	147	147	147	147	147	148	148	148	148	149	149
50	146	147	147	147	148	148	149	149	147	146	146	145	145	146	146	147	148	148	148	149	149
51	146	146	146	147	147	147	147	148	146	145	145	144	144	145	145	146	147	147	148	148	149
52	146	146	146	146	146	146	147	147	146	145	145	145	145	145	145	146	146	146	147	147	148
53	146	146	146	146	146	146	146	146	147	146	146	145	145	146	146	147	145	145	146	146	147
54	147	147	147	147	147	147	147	147	147	147	146	146	146	146	147	147	145	145	146	146	147
55	148	148	148	148	147	147	147	147	146	146	145	145	145	145	146	146	146	146	147	147	148
56	149	149	148	148	148	148	148	147	145	144	144	143	143	144	144	145	147	147	148	148	149
57	149	148	148	147	146	145	145	145	143	143	143	143	143	143	143	143	146	146	146	147	147
58	149	149	148	147	146	146	145	145	143	143	143	143	143	143	143	143	145	146	146	147	147
59	149	149	148	147	146	146	145	145	143	143	143	143	143	143	143	143	145	145	146	146	147
60	148	148	148	147	146	145	145	144	143	143	143	143	143	143	143	142	145	145	145	146	146

The Matrix difference is:

Table 8. Result of the difference between Matrix 1 and Matrix 2

64	Matriu1-Matriu2.																				
65	0	0	0	0	0	0	0	0	0	0	0	1	1	0	0	0	0	0	0	0	0
66	0	0	0	0	0	0	0	0	0	0	0	0	1	0	0	0	0	0	0	0	0
67	0	0	0	0	0	0	0	0	0	1	0	0	0	0	0	0	0	0	0	0	0
68	0	0	0	0	0	0	0	0	0	0	0	0	1	0	0	0	0	0	0	0	0
69	0	0	0	0	0	0	0	0	0	0	0	-1	0	0	0	0	0	0	0	0	0
70	0	0	0	0	0	0	0	0	0	0	0	0	0	0	-1	0	0	0	0	0	0
71	0	0	0	0	0	0	0	0	0	0	0	-1	0	0	0	0	0	0	0	0	0
72	0	0	0	0	0	0	0	0	0	0	0	-1	-1	0	0	0	0	0	0	0	0
73	0	0	0	0	0	0	0	0	0	0	0	0	0	0	0	0	0	0	0	0	0
74	0	0	0	0	0	0	0	0	0	0	0	0	0	0	0	0	0	0	0	0	0
75	0	0	0	0	0	0	0	0	0	0	0	0	0	0	0	0	0	0	0	0	0
76	0	0	0	0	0	0	0	0	0	0	0	0	0	0	0	0	0	-1	0	0	0
77	0	0	0	0	0	0	0	0	0	0	0	0	0	0	0	0	0	0	0	0	0
78	0	0	0	0	0	0	0	0	0	0	0	0	0	0	0	0	0	1	0	0	0
79	0	0	0	0	0	0	0	0	0	0	0	0	0	0	0	0	0	0	1	0	-1
80	0	0	0	0	0	0	0	0	0	0	0	0	0	0	0	0	0	1	0	0	0
81	0	0	0	0	0	0	0	0	0	0	0	0	0	0	0	0	0	0	0	0	0
82	0	0	0	0	0	0	0	0	0	0	0	0	0	0	0	0	0	0	0	0	0
83	0	0	0	0	0	0	0	0	0	0	0	0	0	0	0	0	0	0	0	0	0
84	0	0	0	0	0	0	0	0	0	0	0	0	0	0	0	0	0	0	0	0	0
85	0	0	0	0	0	0	0	0	0	0	0	0	0	0	0	0	0	0	0	0	0
86	0	0	0	0	0	0	0	0	0	0	0	0	0	0	0	0	0	0	0	0	0
87	0	0	0	0	0	0	0	0	0	0	0	0	0	0	0	0	0	0	0	0	0
88	0	0	0	0	0	0	0	0	0	0	0	0	0	0	0	0	0	0	0	0	0
89	0	0	0	0	0	0	0	0	0	0	0	0	0	0	0	0	0	0	0	0	0
90	0	0	0	0	0	0	0	0	0	0	0	0	0	0	0	0	0	0	0	0	0
91	0	0	0	0	0	0	0	0	0	0	0	0	0	0	0	0	0	0	0	0	0
92	0	0	0	0	0	0	0	0	0	0	0	0	0	0	0	0	0	0	0	0	0

The resultant Matrix clearly shows the regions of the image in which all differences have been observed.

2.2. Another Example Has Been Gathered from the Written Press. It Refers to Finding/Spotting the Differences

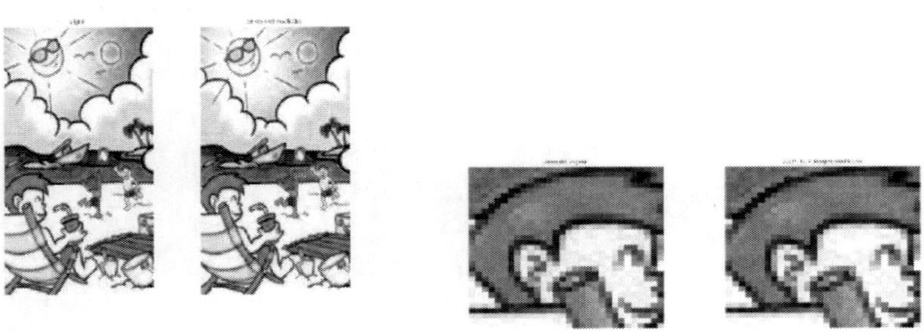

Figure 13. The Game of Differences.

In detail, we realized that as a model, they have respectively their Numerical Matrix:

Table 9. Searching the differences 1st Matrix

105	113	98	103	112	99	99	108	113	115	139	116	74	52	29	38	37	43	54	68	72	70
103	105	92	99	111	106	111	123	128	91	46	32	56	91	114	121	130	128	126	122	122	123
88	86	97	108	99	115	124	77	22	44	72	100	121	130	132	129	128	128	130	131	129	128
98	92	114	96	135	137	27	6	71	94	115	125	125	125	132	140	134	136	137	138	139	138
96	102	86	130	113	19	26	86	118	125	129	128	126	125	128	131	133	131	130	130	130	132
96	99	120	119	23	11	118	135	122	129	133	139	142	138	129	122	128	129	129	129	132	136
108	129	132	28	30	126	118	118	149	151	149	141	133	127	126	125	113	115	118	123	127	129
87	107	38	26	103	117	115	163	141	147	147	142	131	123	118	117	121	118	117	118	123	126
59	0	20	94	110	121	149	131	151	149	145	140	134	129	125	119	91	81	73	66	66	73
237	60	65	111	123	119	135	136	144	122	149	130	117	132	111	47	20	19	25	41	40	32
133	54	114	126	123	127	115	133	132	120	128	130	126	121	35	50	139	193	213	204	199	212
64	90	143	129	118	129	122	132	130	128	112	119	115	46	54	210	240	249	239	247	252	245
76	142	135	126	118	124	133	94	75	77	114	134	112	28	181	236	249	255	244	245	255	252
103	144	120	127	121	127	97	45	44	22	22	49	107	28	198	250	236	255	254	253	254	252
127	122	129	127	120	137	54	65	180	225	176	53	31	42	204	248	255	255	254	255	254	253
130	117	131	123	119	132	34	136	248	212	244	219	57	47	233	236	234	255	255	245	243	255
110	123	117	121	125	114	29	177	231	112	108	226	144	70	233	243	247	231	237	253	243	242
78	126	121	128	115	124	46	177	247	233	148	92	215	168	239	247	244	249	245	234	234	255
41	117	115	115	132	127	35	167	250	170	72	107	243	236	255	214	147	123	120	159	254	243
108	51	130	120	127	130	66	94	255	93	195	214	228	253	163	78	79	93	47	49	243	250
123	37	68	128	118	115	108	33	195	147	120	243	239	154	36	138	191	94	52	66	176	255

Table 10. Searching the differences 2nd Matrix

107	98	120	93	111	91	116	92	102	107	92	105	119	123	120	78	66	49	30	31	41	48	48
105	113	98	103	112	99	99	108	113	115	139	116	74	52	29	38	37	43	54	68	72	70	71
103	105	92	99	111	106	111	123	128	2	46	32	56	91	114	121	130	128	126	122	122	123	124
88	86	97	108	99	115	124	77	22	44	72	100	121	130	132	129	128	128	130	131	129	128	124
98	92	114	96	135	137	27	6	71	94	115	125	125	125	132	140	134	136	137	138	139	138	137
96	102	86	130	113	19	26	86	118	125	129	128	126	125	128	131	133	131	130	130	130	132	133
96	99	120	119	23	11	118	135	122	129	133	139	142	138	129	122	128	129	129	129	132	136	138
108	129	132	28	30	126	118	118	149	151	149	141	133	127	126	125	113	115	118	123	127	129	131
87	107	38	26	103	117	115	163	141	147	147	142	131	123	118	117	121	118	117	118	123	126	131
59	0	20	94	110	121	149	131	151	149	145	140	134	129	125	119	91	81	73	66	66	73	84
237	60	65	111	123	119	135	136	144	122	149	130	117	132	111	47	20	19	25	41	40	32	38
133	54	114	126	123	127	115	133	132	120	128	130	126	121	35	50	139	193	213	204	199	212	216
64	90	143	129	118	129	122	132	130	128	112	119	115	46	54	210	240	249	239	247	252	245	248
76	142	135	126	118	124	133	94	75	77	114	134	112	28	181	236	249	255	244	245	255	252	248
103	144	120	127	121	127	97	45	44	22	22	49	107	28	198	250	236	255	254	253	254	252	255
127	122	129	127	120	137	54	65	180	225	176	53	31	42	204	248	255	255	254	255	254	253	253
130	117	131	123	119	132	34	136	248	212	244	219	57	47	233	236	234	255	255	245	243	255	230
110	123	117	121	125	114	29	177	231	112	108	226	144	70	233	243	247	231	237	253	243	242	221
78	126	121	128	115	124	46	177	247	233	148	92	215	168	239	247	244	249	245	234	234	255	189
41	117	115	115	132	127	35	167	250	170	72	107	243	236	255	214	147	123	120	159	254	243	213
108	51	130	120	127	130	66	94	255	93	195	214	228	253	163	78	79	93	47	240	243	250	240
123	37	68	128	118	115	108	33	195	147	120	243	239	154	36	138	191	94	52	66	176	255	255

As previously shown, the differences between images have been found in row 3, column 10. Over there, number 91 has been converted to 2. Also in row 21, in column 20 number 49 has been converted to 240.

This means that by subtracting both matrices, zero is obtained in all operations except in row 3, column 10, and in row 21, column 20 as well.

To our students, we can comment on another daily example when the Difference between Matrices reaches a remarkable role in the security field.

2.3. The Difference between Matrices as a Model of Security System

According to our previous results, we can compare Images. Think about a hypothetical frame series (in black and white) captured by surveillance camera inside a bank. Now, considering the models of two consecutive images, the "intelligent camera" processes the difference between two Matrices. If Matrix "zero" (null) is obtained, we realize that no movement has been made, in this case it is not necessary to record all associated images to those matrices. On the other hand, all different matrices will be recorded registering the movement inside the bank.

This simplified example explains how the "intelligent surveillance cameras" do an enterprise night surveillance.

3. Shopping at a Supermarket: A Model of the Product between Matrices

In the following situation, the students, naturally discovering how to make the product of matrices into a Model, link the quantities of buying products at a Supermarket to their prices with total expenses.

Now we will do an easy study in order to clarify which of the two supermarkets is the least expensive merchant. We are shopping twice a week, buying the articles and quantities according to the table below:

Table 11. A Daily Consumption

	Pork Loin (kg)	Oranges (kg)	Lettuce (3 units / tray)
1st Day	1	3	1
2nd Day	3	2	2

This table can be written in a different way:

$$Q=\begin{pmatrix} 1 & 3 & 1 \\ 3 & 2 & 2 \end{pmatrix}.$$

And then for each supermarket we calculate:

- What are the expenses for the first day?
- What are the expenses for the second day?

We are also doing a global calculation:

- What is the least expensive option for each day?

These questions could be proposed to our students so that they can calculate and achieve their own conclusions.

First supermarket:

Figure 14. Supermarket 1 Logo.

Figures 15, 16 and 17. Expenses at Supermarket 1.

We are writing the price list in the table below:

Table 12. Articles and Prices in Supermarket 1

Alcampo	Pork Loin (kg)	Oranges (kg)	Lettuce (3 units / tray)
Price	6,49 €	1,25 €	0,75 €

Second supermarket:

Carrefour

Figure 18. 2nd Supermarket Logo.

Figures 19, 20, 21. Expenses at Supermarket 2.

Repeating the same procedure with Carrefour price list:

Table 13. Articles and Prices in Supermarket 2

Carrefour	Pork Loin (kg)	Oranges (kg)	Lettuce (3 units / tray)
Price	5,90 €	0,85 €	1,25 €

Daily Calculation of our expenses in each supermarket:

- Alcampo Supermarket:

1st Day → $1 \cdot 6{,}49 + 3 \cdot 1{,}25 + 1 \cdot 0{,}75 = 10{,}99€$

2nd Day → $3 \cdot 6{,}49 + 2 \cdot 1{,}25 + 2 \cdot 0{,}75 = 23{,}47€$

- Carrefour Supermarket:

1st Day → $1 \cdot 5{,}90 + 3 \cdot 0{,}85 + 1 \cdot 1{,}25 = 9{,}70€$

2nd Day → $3 \cdot 5{,}90 + 2 \cdot 0{,}85 + 2 \cdot 1{,}25 = 21{,}90€$

Mathematically, we can write the expenses in each supermarket as:

– Alcampo Supermarket:

–

1st Day → $(1,3,1) \cdot (6'49,1'25,0'75) = 10'99€$

2nd Day → $(3,2,2) \cdot (6'49,1'25,0'75) = 23'47€$

– Carrefour Supermarket:

–

1st Day → $(1,3,1) \cdot (5'90,0'85,1'25) = 9'70€$

2nd Day → $(3,2,2) \cdot (5'90,0'85,1'25) = 21'90€$

The applied procedure is the well-known *"scaled euclidean product."* It is remarkable that the students had been building the scaled product spontaneously.

Summarizing: A daily matter such as shopping at a Supermarket has a Mathematical Model, the *"scaled euclidean product."*

Globally, we can write the expenses in each supermarket as:

– Alcampo Supermarket 1st and 2nd Day

$$\begin{pmatrix} 1 & 3 & 1 \\ 3 & 2 & 2 \end{pmatrix} \cdot \begin{pmatrix} 6{,}49 \\ 1{,}25 \\ 0{,}75 \end{pmatrix} = \begin{pmatrix} 10{,}99 \\ 23{,}47 \end{pmatrix}$$

– Carrefour Supermarket 1st and 2nd Day

$$\begin{pmatrix} 1 & 3 & 1 \\ 3 & 2 & 2 \end{pmatrix} \cdot \begin{pmatrix} 5{,}90 \\ 0{,}85 \\ 1{,}25 \end{pmatrix} = \begin{pmatrix} 9{,}70 \\ 21{,}90 \end{pmatrix}$$

Here we are naturally building the *product of a matrix by a vector column.* We can join the tables in order to allow a better situation view and the Matrix Model:

– Q: Quantity of Products

Table 14. Articles and Quantities

	Pork Loin (kg)	Oranges (kg)	Lettuce (3 units / tray)
1st Day	1	3	1
2nd Day	3	2	2

$$Q = \begin{pmatrix} 1 & 3 & 1 \\ 3 & 2 & 2 \end{pmatrix}.$$

– P: Price in each supermarket

Table 15. Articles and Prices Supermarket 1

Alcampo	Pork Loin (kg)	Orange s(kg)	Lettuce (3 units / tray)
Price	6,49 €	1,25 €	0,75 €

Table 16. Articles and Prices Supermarket 2

Carrefour	Pork Loin (kg)	Oranges (kg)	Lettuce (3 units / tray)
Price	5,90 €	0,85 €	1,25 €

$$P=\begin{pmatrix} 6{,}49 & 5{,}90 \\ 1{,}25 & 0{,}85 \\ 0'75 & 1{,}25 \end{pmatrix}$$

– D: Expenses Table

Table 17. Daily Expenses in both Supermarkets

	Alcampo	Carrefour
Expenses 1st Day	10,90 €	9,70 €
Expenses 2nd Day	23,47 €	21,90 €

$$D=\begin{pmatrix} 10{,}90 & 9{,}70 \\ 23{,}47 & 21{,}90 \end{pmatrix}$$

By doing it as previously mentioned, it is possible to introduce the *Product of Matrix.*

The following Mathematical Model links Quantities (Q), Price (P), and Expenses (D):

$$\boxed{Q \cdot P = D}$$

Summarizing: We realize that by linking purchased quantities and prices, it is possible to naturally obtain the algorithm to multiply Matrices.

Going back to the previous supermarket comparison, Carrefour Supermarket Chain has the best deals. The total expenses in Carrefour mean significant money saving when compared with those in Alcampo Supermarket.

The students who did this exercise discovered how to multiply Matrices naturally. Now, the professor feels free to propose situations by introducing the inverse matrix concept and other elements of Matrix calculation.

These examples are useful to show how the use of real life situations makes it feasible to find patterns (models) giving information about the proposed situations.

Obviously, the reader can translate the involvement of the Mathematical Modeling in the Academic Curricula, at the same time realizing that competent teaching is taking place.

INDEX

#

A

B

C

D

E

M

N

O

P

Q

R

S

T

U

V

W